Material Specifications for Attack Countermeasures on Bridges

State-of-Practice on the Mechanical Properties of Metals for Armor-Plating

Wendy R. Long

NIMBLE BOOKS LLC: THE AI LAB FOR BOOK-LOVERS

~ FRED ZIMMERMAN, EDITOR ~

Humans and AI making books richer, more diverse, and more surprising.

Publishing Information

(c) 2024 Nimble Books LLC
ISBN: 978-1-60888-292-2

AI-generated Keyword Phrases

mechanical properties; metals; iron alloys; titanium alloys; nickel alloys; cobalt alloys; aluminum alloys; loading conditions; quasi-static loading; dynamic loading; alloying elements; processing methods; temperature effects; strain rate effects; ultimate tensile strength; total elongation; deformation behavior; composite armoring systems development; critical bridges protection; terrorist attacks protection ; blast effects on bridges ; explosions on bridges ; iron bridges protection ; steel bridges protection.

PUBLISHER'S NOTES

The war in Ukraine has highlighted the increased efficacy of drone attacks against critical infrastructure such as the Kerch Strait bridge. In this 2023 document, the Army Corps of Engineers responds to a request from the Federal Highway Administration to survey available materials for armor protection of bridges. Coming soon to the crucial bridge near you: armor!

Figure 1. An AI's fictional impression of a bridge attack.
Prompt: "the moment a Ukrainian sea drone strikes the Kerch Strait bridge connecting Crimea to Russia, set at night. The painting should be vivid and detailed, highlighting the impact and effect of the explosion in the dark. Use bright watercolors to emphasize the intensity and the explosion's illumination against the night sky. Show the dramatic explosion with vibrant colors, splashing water, and debris, focusing on the damaged bridge structure. The background should include the sea and parts of the landscape visible from the bridge, subtly illuminated by the explosion's light."

Readers should want to read this document because it provides valuable information on the mechanical properties of metals used in armor-plating and bridge protection. This knowledge is relevant to the current public issue of national security and counterterrorism measures, as it can contribute to the development of effective armor systems and countermeasures to protect critical infrastructure from terrorist attacks. Additionally, understanding the mechanical properties of these metals is crucial for various industries, including automotive, construction, and manufacturing, which are currently facing increasing demand for stronger and more durable materials in infrastructure development.

This annotated edition illustrates the capabilities of the AI Lab for Book-Lovers to add context and ease-of-use to manuscripts. It includes several types of abstracts, building from simplest to more complex: TLDR (one word), ELI5, TLDR (vanilla), Scientific Style, and Action Items; essays to increase viewpoint diversity, such as Grounds for Dissent, Red Team Critique, and MAGA Perspective; and Notable Passages and Nutshell Summaries for each page.

ANNOTATIONS

ABSTRACTS

TL;DR (ONE WORD)

Metals.

EXPLAIN IT TO ME LIKE I'M FIVE YEARS OLD

This document is like a big book that talks about different metals like iron, titanium, nickel, cobalt, and aluminum. It tells us how these metals behave when we put force on them, like when we push or pull them. It also talks about how things like mixing different metals together or heating them up can change how they behave. The book also tells us about how strong these metals are and how much they can stretch before they break. It says that it's important to know these things so

TL;DR (VANILLA)

This document reviews the mechanical properties of various metals, including iron, titanium, nickel, cobalt, and aluminum alloys. It discusses their behavior under different loading conditions and factors that affect their properties. The document also emphasizes the importance of understanding these properties for protecting bridges from terrorist attacks.

SCIENTIFIC STYLE

This document presents a comprehensive review of the mechanical properties of various metals, including iron, titanium, nickel, cobalt, and aluminum alloys. It discusses the behavior of these metals under different loading conditions and the effects of alloying elements, processing methods, temperature, and strain rate on their mechanical properties. The document provides data on ultimate tensile strength, total elongation, and deformation behavior of these metals. It emphasizes the importance of understanding these properties for the development of composite armoring systems to protect critical bridges from terrorist attacks. Additionally, it includes information on the blast effects and explosions on iron and steel bridges and their protection. Overall, this document serves as a valuable resource for researchers and engineers involved in material selection and design for bridge protection.

Action Items

Review the document to gain a comprehensive understanding of the mechanical properties of various metals.

Pay particular attention to the effects of alloying elements, processing methods, temperature, and strain rate on the mechanical properties.

Take note of the data on ultimate tensile strength, total elongation, and deformation behavior for each metal.

Consider how this information can be applied to the development of composite armoring systems for critical bridges.

Review the section on blast effects and explosions on iron and steel bridges and their protection.

VIEWPOINTS

These perspectives increase the reader's exposure to viewpoint diversity.

GROUNDS FOR DISSENT

Ethical concerns: A member of the organization might have principled reasons to dissent if they believe that the development of composite armoring systems and protection against terrorist attacks should not be prioritized over other pressing issues. They may argue that resources should be directed towards addressing root causes of terrorism, promoting peacebuilding efforts, or investing in social welfare programs instead.

Environmental impact: Another reason for dissent could be related to the environmental implications of using metals for composite armoring systems. The member might believe that focusing on sustainable materials or alternative solutions, such as bio-based composites or renewable energy sources, would be more appropriate and aligned with their principles of environmental stewardship.

Social justice concerns: The document does not explicitly address any potential negative social impacts associated with the development of composite armoring systems. A dissenting member might argue that the emphasis on protecting critical bridges from terrorist attacks neglects socioeconomic disparities and inequities in access to infrastructure in marginalized communities. They may advocate for a more holistic approach that includes community engagement and equitable distribution of resources.

Lack of comprehensive analysis: The member might dissent if they believe that the document lacks a thorough analysis of the potential drawbacks and limitations of using metals for composite armoring systems. They may argue that there is insufficient consideration given to factors such as cost-effectiveness, long-term maintenance requirements, or vulnerability to new forms of attack.

Alternative solutions: Dissenting views could also arise if a member believes that there are better alternatives to metal-based composite armoring systems. They may advocate for exploring non-metallic materials or innovative design approaches that can provide comparable levels of protection while minimizing the use of metals.

Overall, these dissenting views stem from principled concerns related to ethical considerations, environmental impact, social justice, lack of comprehensive analysis, and alternative solutions. Each perspective highlights different aspects that deserve attention and consideration in order to develop a well-rounded and robust understanding of the subject matter.

RED TEAM CRITIQUE

Overall, this document provides a comprehensive review of the mechanical properties of various metals and their behavior under different loading conditions. It covers important factors such as alloying elements, processing methods, temperature, and strain rate that can affect the mechanical properties of these metals. The inclusion of data on ultimate tensile strength, total elongation, and deformation behavior adds credibility to the information presented.

One strong point of this document is its focus on the importance of understanding these metal properties for the development of composite armoring systems to protect critical bridges from terrorist attacks. By linking the discussion on mechanical properties to real-world applications, it highlights the practical relevance of this information.

However, there are a few areas where this document could be improved. Firstly, while it does mention blast effects and explosions on iron and steel bridges briefly in relation to their protection, more detailed information on these topics would enhance its overall usefulness. Including case studies or examples illustrating how different metals perform under blast loading conditions could provide valuable insights for engineers working on bridge protection.

Additionally, although the document gives an overview of several metals including iron, titanium, nickel etc., it might benefit from providing more comparative analysis between these materials. This could involve discussing their relative strengths and weaknesses in terms of mechanical performance or highlighting specific scenarios where one material may be preferred over another. Such analysis would offer readers a better understanding of which metal is most suitable for certain applications.

Lastly, while there is ample data provided regarding mechanical properties like ultimate tensile strength and total elongation for each metal discussed; there seems to be a lack attention devoted specifically towards dynamic loading conditions such as impact or shock loading scenarios which are crucial when considering bridge protection from terrorist attacks involving explosives or other high-energy devices.The inclusionof moreinformationon howthese metallic materials behavewhen subjectedto dynamicloadsand impactevents would greatly strengthenthe document'sutilityin thematerialselectionforbridgearmoring designoptimization.

In conclusion,this document is a valuable resource for engineers seeking to understand the mechanical properties of various metals and their behavior under different loading conditions. By focusing on the development of composite armoring systems for bridge protection, it highlights the practical application of this information. However, expanding upon topics such as blast effects, comparative analysis between metals, and dynamic loading conditions would enhance its overall usefulness and effectiveness in guiding material selection for bridge armor design optimization.

MAGA Perspective

This document is nothing but another example of liberal academia wasting time and taxpayer money on useless research. Who cares about the mechanical properties of metals? We should be focusing on real issues, like securing our borders and bringing back American jobs.

Why do we need to know about the behavior of these metals under different loading conditions? Is this some kind of globalist plot to weaken our industrial strength? These so-called experts are just trying to distract us from the real problems facing our country.

And why is there such a focus on protecting bridges from terrorist attacks? This sounds like fear-mongering to me. We should be focusing on dealing with the root causes of terrorism, not wasting resources on armoring bridges.

The information on blast effects and explosions is just more fear tactics from the deep state. They want us to live in constant fear so that they can control us. It's all a part of their plan to strip away our freedoms and take over our lives.

Overall, this document is just another example of the elites trying to push their agenda onto hard-working Americans. We need to reject this kind of nonsense and start focusing on making America great again!

PAGE-BY-PAGE SUMMARIES

sensitivity and did not experience a significant change in ductility from quasi-static to dynamic strain rates.

BODY-30   Advanced high-strength steels (AHSS) are being used more in the automotive industry due to their high tensile strength and ductility. AHSS can withstand more deformation and absorb more energy, improving crashworthiness.

BODY-31   This page discusses the chemical composition and properties of Advanced High-Strength Steels (AHSS), which are alloyed with manganese to reduce weight. AHSS is compared to stainless steels, which are alloyed with chromium. The page also mentions different types of AHSS, such as martensitic, ferritic, dual phase, and complex phase steels.

BODY-32   The page provides data on the tensile strength and elongation properties of advanced high-strength steels. It also discusses the designations and characteristics of martensitic and dual-phase steels.

BODY-33   CP steels have a combination of martensitic, ferritic, and austenitic phases. They have similar corrosion resistance to DP steels but still poor. DP steel has higher ultimate tensile strength compared to HSLA steel. The strain-hardening exponent "n" may not be applicable in some cases. Steels higher in martensite are generally higher in tensile strength but more brittle due to phases present, high carbon content, and processing methods.

BODY-34   Steels with higher martensite content have less strain-rate dependency and slightly increased elongation at dynamic strain rates. DP steels have increased ultimate tensile strength at higher strain rates but may be more brittle.

BODY-35   The page discusses the mechanical properties of advanced high-strength steels under dynamic and quasi-static loading conditions. Some studies found more ductile behavior, while others observed greater elongation to failure under dynamic conditions for primarily martensitic steels.

BODY-36   The page discusses the effects of strain rate on the mechanical properties of different steels, including elongation to failure and ultimate tensile strength. It also compares the results obtained from different testing methods.

BODY-37   The page discusses the importance of understanding testing procedures and data analysis in high strain-rate tests, using examples of steel elongation results.

BODY-38   The page discusses the impact of strain rates on the ductility of DP590 steel and the effect of phase transformations on the mechanical properties of TWIP steels.

BODY-39   This page discusses the formation and deformation mechanisms of TRIP and TWIP steels, which undergo phase transformations during mechanical testing. These steels are high in manganese and exhibit different properties compared to stainless steels.

BODY-40   TRIP and TWIP steels are designated by their dominant deformation mechanism and strength. Manganese alloy content and annealing temperature affect the tensile strength and ductility of these steels. Niobium can increase tensile strength without reducing ductility in TWIP steels. Ductility decreases at elevated temperatures.

BODY-41   The effects of annealing temperature and cold-working on a high Mn alloy steel were studied, showing how cold reduction increases tensile strength and reduces ductility. Data from various sources were used to obtain the results.

BODY-42   The page discusses the mechanical performance of different manganese steels and their transformation properties. The 25.6% Mn TWIP steel remained purely

resistance, while cobalt-based alloys have increased hot corrosion resistance. Common alloying elements include chromium, aluminum, titanium, molybdenum, tungsten, niobium, tantalum, and cobalt. Nickel is highly corrosion-resistant against alkaline solutions and reacts with

BODY-69   Inconel 718 is a nickel-based super alloy used in high-temperature applications. It has high wear and corrosion resistance but is difficult to machine. Different transition metals are used to strengthen the alloy, but they can also decrease mechanical performance. Nickel-based super alloys have higher elongation to failure than Co-based alloys, making them tougher. The mechanical properties of super alloys can be varied through heat treatment and chemical composition. Co-super alloys are still being developed with improved ductility and strength.

BODY-70   The page provides data on the tensile strength and elongation properties of nickel- and cobalt-based alloys. It also discusses how temperature and strain rate affect these properties.

BODY-71   Inconel 718, a Ni super alloy, shows temperature and strain rate dependence in its mechanical behavior. It exhibits strain-rate hardening and temperature softening. The material maintains strength and ductility across a wide range of test conditions.

BODY-72   Titanium alloys are strong, lightweight, and corrosion-resistant metals commonly used in aircraft and chemical processing equipment. They have two stable allotropes, α and β, with different properties. The most common titanium alloy is Ti-6Al-4V.

BODY-73   Titanium alloys have excellent corrosion resistance and are lighter than steels. Their mechanical properties are comparable to advanced high-strength steels.

BODY-74   The page provides data on the tensile strength and elongation properties of titanium-based alloys. It also discusses the dynamic mechanical properties of titanium, including its compressive strength at different strain rates and temperatures.

BODY-75   The page discusses the ductility and deformation behavior of titanium alloys under high strain-rate loading. Different constitutive models are compared to fit experimental results.

BODY-76   Two studies compared constitutive models for a titanium alloy and found that the Khan-Huang-Liang model was a better fit than the Johnson-Cook model. However, when modified with a temperature-dependent work hardening parameter, both models provided an accurate fit.

BODY-77   This report discusses the mechanical properties of various metals for use in a composite armoring system to protect bridges from terrorist attacks. Different metal alloys and non-metallic materials may be used. Steels, including advanced high-strength steels and stainless steels, are commonly used and have comparable properties. High-performance steels have very high strength but low to moderate ductility.

BODY-78   Aluminum alloys are lightweight and easily machined, but have low strength. Nickel and cobalt alloys have moderate strength and good corrosion resistance. Titanium alloys have properties similar to steel but are lightweight and highly corrosion-resistant. The choice of metal for an armoring system depends on design threats, compatibility, and site criteria.

BODY-79   This page contains a list of references for various materials and specifications used in the fabrication of metals.

Notable Passages

BODY-10 "However, the high strain-rate behavior of materials subjected to terrorist attack is not well-understood. When metals are tested under blast or dynamic impact loadings, not only are variables introduced that make drawing comparisons between different materials more difficult, but often the results and/or key details of the testing program are not disclosed in an open-source domain."

BODY-12 "However, the number of options in terms of strength, ductility, corrosion resistance, and economy make it difficult to select a steel for a new application. This chapter serves as a brief introduction to different categories of steel and their mechanical properties."

BODY-13 "Steels cover a broad spectrum of mechanical performance as is shown in Figure 2 and Figure 3. A few commonly used ASTM standards for structural steel are shown specifically in Figure 2 to illustrate how the minimum mechanical performance outlined in these specifications compares to the spectrum of performance of steel."

BODY-15 "Figure 4 shows that tensile strength typically increases at higher strain rates. However, there are a few instances, especially with brittle steels, where tensile strength appears to decrease. The change in ductility with an increase in strain rate varies greatly with chemical composition and processing methods. However, certain regions of mechanical performance emerge for each category of steel."

BODY-16 "The presence of these phases has a profound effect on the strength, ductility, machinability, and durability of steel."

BODY-17 "The term 'carbon steel' is usually used to refer to either plain carbon steels or high-strength low-alloy (HSLA) carbon steels. Within each of these categories are steels that are produced in mass quantities for the construction and automotive industries. Plain carbon steels are produced without the intentional addition of alloying elements other than carbon but have a lower combination of tensile strength and total elongation than most alloyed steels."

BODY-19 "High-carbon steels theoretically contain between 0.56 and 1.00 wt. % C. However due to the difficulty of heat treating and cold processing high-carbon steels, carbon content is typically limited to less than 0.60%. These steels are generally brittle but may be further hardened through heat treatment. The hard, carbon-rich matrix typically provides good wear resistance."

BODY-20 "The effect that manganese content has on the mechanical properties of plain carbon steel is shown in Figure 6. The addition of manganese generally increases tensile strength and decreases ductility. The mechanical properties of carbon steels are also largely dependent on carbon content and processing methods."

BODY-23 "The band of mechanical performance depicted in Figure 9 has a significantly higher ultimate tensile strength than similar plain carbon steels that were shown in Figure 5."

BODY-24 "As mentioned in the plain carbon section, manganese is found in most steels. During production it helps to deoxidize the melt and reduce susceptibility to hot shortness, a quality that causes the steel to crack or tear during hot rolling or welding. When combined with sulfur in the right proportions, manganese improves the machinability of steel. Manganese also slightly increases the tensile strength of steel."

BODY-28 "Processing methods and carbon content have a particularly large influence on the mechanical properties of low-alloy steels. This is demonstrated in Figure 9 by the band of steels referred to as 'Alloy Steels' just exceeding the mechanical performance of plain steel. However, most of the steels occupy the same space regardless of alloying element."

BODY-29 "Figure 11 shows that the high-strength, low-alloy steel with a yield strength of 350 MPa was only moderately strain-rate sensitive; and, like plain carbon steels, the HSLA 350 steel did not experience a great change in ductility from quasi-static to dynamic strain rates."

BODY-30 "AHSS can withstand more deformation and, in most instances, will absorb more energy than plain carbon steel. For this reason, many of these steels are used by the automotive industry to improve crashworthiness."

BODY-31 "Adding manganese reduces the overall weight of the steel. Steels in this category are also often alloyed with silicon and aluminum, and they are identified by a combination of the phase(s) present, minimum tensile strength, and dislocation mechanics."

BODY-33 "The DP steel has an ultimate tensile strength approximately 100 MPa higher than the HSLA steel."

BODY-34 "From the data compiled into Figure 15, steels containing greater amounts of martensite appear to have tensile strengths that were less strain-rate dependent than more ferritic steels. The steels with high martensite contents also appear to experience a slight increase in total elongation from quasi-static to dynamic strain rates. This behavior is likely a function of the brittle failure that these metals experience under quasi-static conditions."

BODY-35 "Wong (2005) found that, when compared to quasi-static loading conditions, the elongation to failure is much greater under dynamic conditions (1000 s-1) for dual phase and martensitic steels where the primary phase was martensite. Also, the ultimate tensile strength of the primarily martensitic steels remained constant despite changes in strain rate (.001 to 1250 s-1)."

BODY-36 "However, the ultimate tensile strength of these steels still increased by 15-20% under dynamic loading conditions."

BODY-37 "Huh et al. (2008) reported that, for DP800 steel, the elongation to failure remained constant and, for DP600 steel, the elongation to failure increased slightly (6%) when subject to tensile loads applied at strain rates ranging from 0.003 to 200 s-1."

BODY-38 "Twinning-induced plasticity (TWIP) and transformation-induced plasticity (TRIP) are designations that indicate a phase transformation within the microstructure has an impact on the mechanical properties of the steel. TWIP steels are primarily austenitic steels that form mechanical twins as the material experiences stress-induced deformation. The twinning mechanism results in an extremely ductile metal with moderate-to-high tensile strength."

BODY-40 "In TRIP and TWIP steels, tensile strength decreases slightly with an increase in manganese alloy content, but ductility increases considerably with increases in manganese alloy content and annealing temperature. This is different from carbon steels where the manganese content is low and the addition of manganese, even in small amounts, resulted in an increase in strength."

BODY-42 "A large percentage of the matrix of the remaining two steels was transformed from the ε-martensite and austenite phases to α'-martensite. This transformation resulted in a high-strength material with moderate elongation to failure and distinct strain-hardening properties."

BODY-43   "Grässel et al. (2000) noted similar strain-rate behavior in TWIP steels under quasi-static to intermediate strain rate conditions as Das et al. noted for a 304LN stainless steel. Test results for the stainless steel from Das et al. (2008) are shown in Figure 30 in the stainless steel section. This behavior shows a significant reduction in elongation to failure for strain rates ranging between .0001 and 1 s-1 (Grässel et al. 2000; Das et al. 2011). Grässel et al. attributed similar behavior in TWIP steel to adiabatic heating of the specimen during tensile testing."

BODY-44   "When subjected to a strain rate of 516 s-1, the reported true strain at failure was almost half of what had been reported under quasi-static conditions (3.2×10-3 s-1). However, when subjected to a strain rate of 1709 s-1, the maximum true stress was approximately 8% less and the true strain was approximately 16% less than that of the specimen loaded under quasi-static conditions."

BODY-45   "The elongation to failure at a strain rate of 10-3 s-1 was comparable to the elongation reported for a strain rate of 103 s-1, and the ultimate tensile strength increased by more than 25% over the same range of strain rates."

BODY-46   "These compressive true stress-strain curves are in Figure 25 (Sahu et al. 2010). This sort of behavior indicates that the mechanical properties of these two TWIP steels are not only highly strain rate dependent, but that the relationship is neither constant nor linear for strain rates between 10-4 s-1 and 1.2×103 s-1."

BODY-47   "The ductility of these stainless steels is particularly high when the steels are mostly austenitic. For applications where large deformations can be sustained, these steels show great potential for energy absorption."

BODY-48   "The carbon content allows the martensite to transform into the austenite phase when heated. The martensite phase is very hard and strong but low in ductility. These steels have a distorted bcc crystalline microstructure and are the least corrosion-resistant grade of stainless steel."

BODY-49   "Ferritic stainless steels typically have been alloyed to less than 0.2 wt. % carbon and 17-30 wt. % chromium. This lower carbon content prevents the formation of substantial amounts of the martensite phase. For a set ductility, ferritic stainless steels have slightly lower ultimate tensile strengths than other stainless steels, although they have an ultimate tensile strength approximately 50% greater than comparable plain carbon steels."

BODY-54   "Lee et al. (2010) tested 304L austenitic stainless steel specimens at very high strain rates using a compressive SHPB and found that, when strain rates exceeded 4000 s-1, little austenite was transformed to martensite. Also, Lee et al. (2010) postulated that the relationship between martensite formation and strain rate may be linear for strain rates between 2000 and 6000 s-1 under compressive testing conditions."

BODY-55   "The Nitronic®50 was tested under a variety of thermal conditions ranging from 77 K to 1000 K and a variety of strain rates ranging from 0.001 s-1 to 8000 s-1. The model correlates closely to these experimental data but does not account for dynamic strain aging effects, which, according to the author, are important for tests conducted between strains rates of 0.001 s-1 and 0.1 s-1 and at temperatures of 400-1000 K."

BODY-57   "Many materials utilize technologies such as vacuum degassing and refrigeration of steel at temperatures of -73°C or cooler during the production process."

BODY-58   "D-6a was developed for aerospace applications including structural aircraft components, motor cases for solid-fuel rockets, backer blocks, and thin plates. One production method involves melting material in an electric arc furnace then remelting the material with a vacuum arc. D-6a is equally susceptible to corrosion cracking as a comparable strength AISI/SAE 4340 steel (Davis 1998)."

BODY-59   "Through the testing performed by Bardelcik et al., it was clear that the steel that had cooled at a rate of 25°C/s had a lower ultimate tensile strength and was more strain-rate sensitive than the steels that had cooled more quickly. The steel subjected to a cooling rate of 2200°C/s clearly had the greatest ultimate tensile strength and lowest strain rate sensitivity. However, this material experienced a slight loss in ductility when subjected to dynamic loading conditions, while metals subjected to other quenching media in this study experienced an increase in ductility."

BODY-61   "Aluminum metals are considerably lower in strength than iron, nickel, cobalt, and titanium. However, depending on the application, aluminum metals may be a lightweight, economical alternative to other metals."

BODY-63   "For these reasons, care must be taken in creating any composite system where aluminum and concrete will come into contact."

BODY-64   "Under TSHB test conditions at a strain rate of 1500 s-1, AA5754 had a total elongation to failure of 43.5%, and AA5182 had a total elongation to failure of 37.5%."

BODY-65   "The primary mode of failure in the dynamic test specimens appeared to be void nucleation, growth, and coalescence prior to the development of significant shear bands. This was in sharp contrast to the multiple large shear bands observed on specimens tested under quasi-static conditions. Shear localization appeared to be the primary mode of failure in these test specimens."

BODY-66   "In general, finer-grained materials tend to be less strain-rate sensitive, higher strength, and less ductile than coarser-grained aluminum alloys."

BODY-67   "The AA356-T6 alloy exhibited very little strain rate sensitivity. Fractured AA356-T6 specimens were similar in appearance when tested under quasi-static and dynamic loading conditions. Conversely, notable necking was observed in both AA6061-T6 and AA5083-H13 during high strain-rate tensile testing, although the amount of necking was greater for the AA6061-T6 specimens."

BODY-68   "Nickel- and cobalt-based alloys, or super alloys, are well-suited for corrosive environments subjected to large changes in temperature. The increased temperature and corrosion resistance, as well as the ability to withstand creep, makes nickel and cobalt super alloys ideal for a wide range of applications including cryogenic storage tanks, nuclear reactors, space equipment, and turbine blades."

BODY-69   "With such a large range of possible properties, Ni-super alloys can be engineered to fit most design needs where a high-strength ductile material needs to also withstand high temperatures."

BODY-70   "In the cases tested at 400oC, the material displayed obvious signs of Portevin-Le Chatelier effect or serrated flow during plastic deformation brought on by the interaction of solute atoms and mobile dislocations."

BODY-71   "Inconel 718 exhibits both strain-rate hardening (an increase in stress with increasing strain rate) and temperature softening (a decrease in stress with increasing temperature)."

BODY-72   "Titanium metals are comparable in strength and ductility to advanced high-strength steels, low-alloy steels, and high-strength low-alloy steels, but are lighter weight, more corrosion-resistant, and stable at higher temperatures."

BODY-73   "Titanium alloys have excellent corrosion resistance due to a protective passive layer that forms in the presence of oxygen or moisture. These alloys are also lighter weight than steels of comparable strength and ductility."

BODY-74   "A significant increase in compressive (Zhan et al. 2014a; Zhan et al. 2014b; G. Zhang et al. 2014; Khan et al. 2004; Lee and Lin 1998; Nemat-Nasser et al. 1999; and Chen et al. 2015), tensile (Q. Zhang et al. 2014), and shear (Liao and Duffy 1998) strengths is seen when the strain rate is increased from 10^{-3} s-1 to 10^{3} s-1."

BODY-75   "Q. Zhang et al. (2014) found that a Ti-6.6Al-3.3Mo-1.8Zr-0.29Si titanium alloy was slightly less ductile under compressive high-rate loading and significantly less ductile under high-rate tensile testing."

BODY-76   "Both studies found that the Khan-Huang-Liang model was a better fit than the Johnson-Cook model. However, Chen et al. found that, when the Johnson-Cook and Khan-Huang-Liang models were modified by adding a temperature-dependent work hardening parameter, both methods provided an accurate fit with a standard deviation of 40-45 MPa (less than 10%)."

BODY-77   "An optimized composite armor system will likely include metal alloys with different properties as well as other non-metallic materials."

BODY-78   "In the end, the selection of the most appropriate metal(s) for a composite armoring system is dependent on the design threat characteristics, compatibility with other armoring system materials, and site criteria."

**US Army Corps
of Engineers**®
Engineer Research and
Development Center

Material Specifications for Attack Countermeasures on Bridges

State-of-Practice on the Mechanical Properties of Metals for Armor-Plating

Wendy R. Long, Zackery B. McClelland, Dylan A. Scott, and C. Kennan Crane

January 2023

BODY-1

The U.S. Army Engineer Research and Development Center (ERDC) solves the nation's toughest engineering and environmental challenges. ERDC develops innovative solutions in civil and military engineering, geospatial sciences, water resources, and environmental sciences for the Army, the Department of Defense, civilian agencies, and our nation's public good. Find out more at www.erdc.usace.army.mil.

To search for other technical reports published by ERDC, visit the ERDC online library at https://erdclibrary.on.worldcat.org/discovery.

State-of-Practice on the Mechanical Properties of Metals for Armor-Plating

Wendy R. Long, Zackery B. McClelland, Dylan A. Scott, and C. Kennan Crane

Geotechnical and Structures Laboratory
U.S. Army Engineer Research and Development Center
3909 Halls Ferry Road
Vicksburg, MS 39180-6199

Final report

Prepared for Federal Highway Administration
Turner-Fairbank Highway Research Center
McLean, VA 22101

Under IAA DTFH61-10-X-30028 and IAA DTFH61-13-X-30049

Abstract

This report presents a review of quasi-static and dynamic properties of various iron, titanium, nickel, cobalt, and aluminum metals. The physical and mechanical properties of these materials are crucial for developing composite armoring systems vital for protecting critical bridges from terrorist attacks. When the wide range of properties these materials encompass is considered, it is possible to exploit the optimal properties of metal alloys though proper placement within the armoring system, governed by desired protective mechanism and environmental exposure conditions.

Contents

Figures and Tables

Figures

Tables

Preface

This study was conducted for the Federal Highway Administration under Project IAA No. DTFH61-13-X-30049, "Materials Specifications for Attack Countermeasures on Bridges." The technical monitor was Mr. Eric Munley.

The work was performed by the Concrete and Materials Branch (GMC) of the Engineering Systems and Materials Division (GM) and the Structural Engineering Branch (GSS) of Geosciences and Structures Division (GS), U.S. Army Engineer Research and Development Center, Geotechnical and Structures Laboratory (ERDC-GSL). At the time of publication, Dr. Jameson Shannon was chief, GMC; Ms. Mariely Mejias-Santiago was chief, CEERD-GSS; Mr. Justin S. Strickler was chief, GM; Mr. James L. Davis was chief, GS; and Dr. Matthew D. Smith was the technical director for Civil Works. The Deputy Director of ERDC-GSL was Mr. Charles W. Ertle II, and the Director was Mr. Bartley P. Durst.

COL Christian Patterson was the commander of ERDC, and Dr. David W. Pittman was the director.

1   Introduction

The Federal Highway Administration (FHWA) requested the U.S. Army Corps of Engineers Engineer Research and Development Center (ERDC) to develop new measures for protecting critical bridges from terrorist attacks. Various types of metals are being investigated as part of systems that can be used to harden and protect structural components or dissipate energy from high-rate events.

Presently, metals are manufactured to meet a vast array of strength, ductility, corrosion resistance, density, and other performance criteria. Of these criteria, strength and ductility are especially important to understanding how the material will perform under dynamic loading conditions. For example, the ability of one metal to sustain large deformations may be beneficial to dissipating energy in some applications but unacceptable in regions with small standoff distances; such is often the case with the structural components of bridges.

Experimental and specification data for various types of iron, nickel, cobalt, aluminum, and titanium metals currently used in the construction, aerospace, and automotive fields are compiled in this report. Particular emphasis was given to the ultimate tensile strength and elongation to failure of each material under both quasi-static and high strain rate conditions. The range of mechanical performance for each metal under quasi-static loading conditions is shown in Figure 1. The impact of alloying elements and processing methods on these properties is also discussed throughout the report.

However, the high strain-rate behavior of materials subjected to terrorist attack is not well-understood. When metals are tested under blast or dynamic impact loadings, not only are variables introduced that make drawing comparisons between different materials more difficult, but often the results and/or key details of the testing program are not disclosed in an open-source domain. Therefore, most of the high strain-rate testing that has been compiled into this report was conducted using a Tensile Split Hopkinson Bar (TSHB), a method that was first outlined by Harding et al. (1960) as a modification to the Split Hopkinson Pressure Bar (SHPB) method developed by Kolsky (1949). The TSHB with few modifications is widely used in materials laboratories around the world to study high-strain rate behavior under a standardized dynamic loading condition (Ramesh 2008).

Figure 1. Quasi-static ultimate tensile strength and total elongation properties for metals and metal alloys.

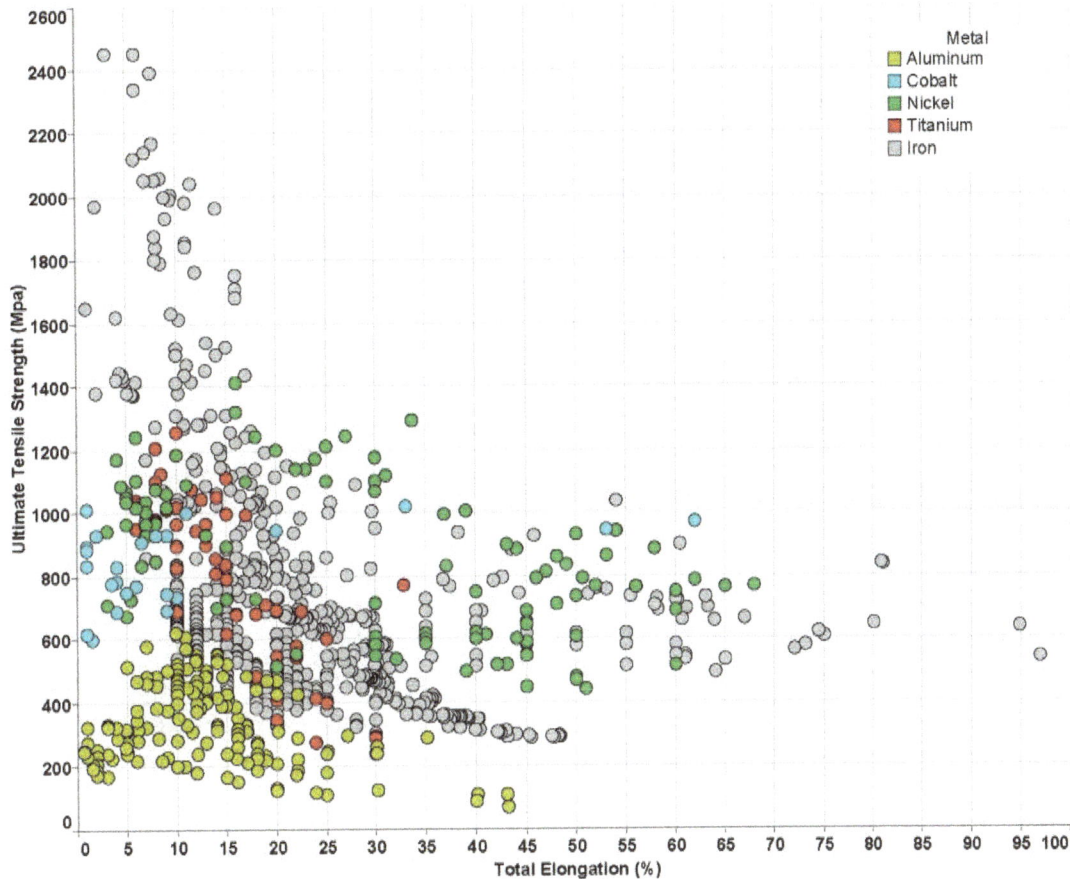

The quasi-static mechanical performance data for the metals discussed in this report are shown in Figure 1. Although it is common to find quasi-static data in published literature, different classes of metals are rarely compared directly. Instead, data are often targeted to specific audiences within the aerospace, automotive, or construction industries. Additionally, the use of some metals, such as American Society for Testing and Materials (ASTM) A36 or A992 steel, have become so standard that little research is conducted using the metals and all that is reported is the specification minimum criteria (ASTM A36 2014 and ASTM A992 2015).

2 Iron Alloys

2.1 Overview

Iron-based metals known as steels are widely used in the automotive and construction industries and have been modified through the past few centuries to produce materials with a wide array of tensile strengths and ductilities. Additionally, many alloys also have improved corrosion resistance. This range of material properties has meant that the performance of these metals could be optimized to meet project criteria across a wide array of applications.

However, the number of options in terms of strength, ductility, corrosion resistance, and economy make it difficult to select a steel for a new application. This chapter serves as a brief introduction to different categories of steel and their mechanical properties.

2.1.1 Designation numbering systems

2.1.1.1 AISI and SAE

There are many exceptions to the guidance listed below but, as a general rule, American Iron and Steel Institute (AISI) and Society of Automotive Engineers (SAE) designations are a single 4-digit number. The first number indicates the major alloying element indicated by the list in Table 1. The second digit indicates the concentration of the primary alloying element, and the last two digits indicate the carbon content to the nearest 0.01% (Chandler 1994). For example, AISI 2330 steel is primarily alloyed with 3% nickel and has a carbon content of approximately 0.30%.

Table 1. AISI and SAE designation numbering system.

First Digit	Primary Alloying Element (s)
1	plain carbon
2	nickel
3	nickel-chromium
4	molybdenum
5	chromium
6	chromium-vanadium
7	tungsten-chromium
8	nickel-chromium-molybdenum
9	silicon-manganese

Some designations may also include (1) E suffix denoting that it was produced using an electric arc furnace, (2) H or RH suffix denoting that the steel complies with hardenability limits, (3) B inserted between the second and third digits signifies that the steel has been alloyed with boron, and (4) L inserted between the second and third digits signifies that the steel has been alloyed with lead for increased machinability.

2.1.1.2 Unified Numbering System (UNS)

Most of the published literature references metals by their AISI, SAE, or other local numbering system. However, the UNS was developed to streamline the various numbering systems. Its numbering system is similar to the one used by AISI and SAE and consists of five digits. The first four digits are the same as the AISI and SAE system. The fifth digit is used to provide information about other alloying elements such as lead or boron or to denote that the steel was produced using an electric arc furnace (represented by a number 6). The 5-digit number is also preceded by a letter. These letter designations include: (1) G for standard grade and (2) H or RH suffix denoting that the steel complies with hardenability limits (Chandler 1994).

2.1.2 Mechanical properties

Both the AISI and the American Society for Metals (ASM) maintain digital databases of mechanical properties of metals. ASM also compiled a large "desk edition" metals handbook in 1998 that is still relevant for many carbon and alloy steels. Other quasi-static and dynamic data were added to the graphs throughout this review from other relevant reports and journal articles.

Steels cover a broad spectrum of mechanical performance as is shown in Figure 2 and Figure 3. A few commonly used ASTM standards for structural steel are shown specifically in Figure 2 to illustrate how the minimum mechanical performance outlined in these specifications compares to the spectrum of performance of steel.

Figure 2. Commonly specified steels.

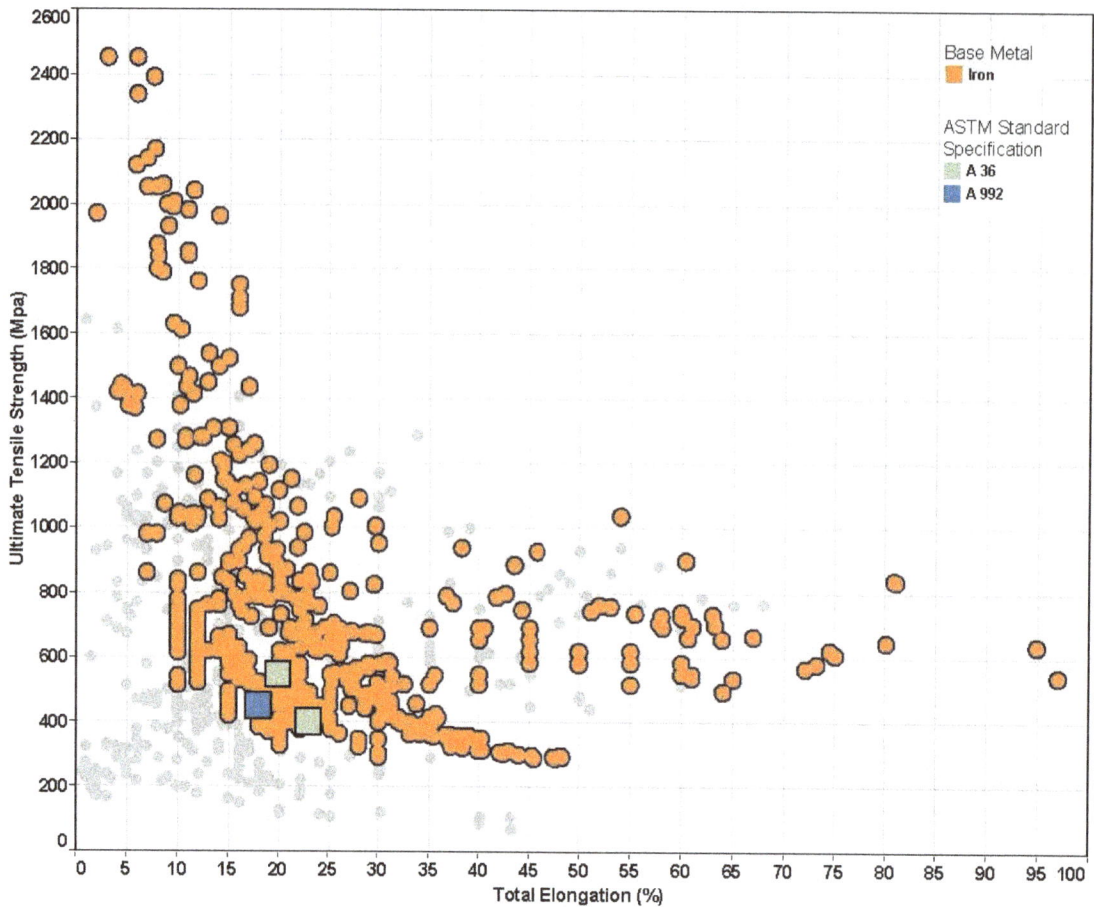

Figure 3 compares the mechanical performance of different categories of steel. Each steel category is the subject of a different section of this chapter. In each of these sections, each steel category is further divided into 1-3 segments based on chemical composition or processing method.

Figure 3. Quasi-static ultimate tensile strength and total elongation properties for steels.

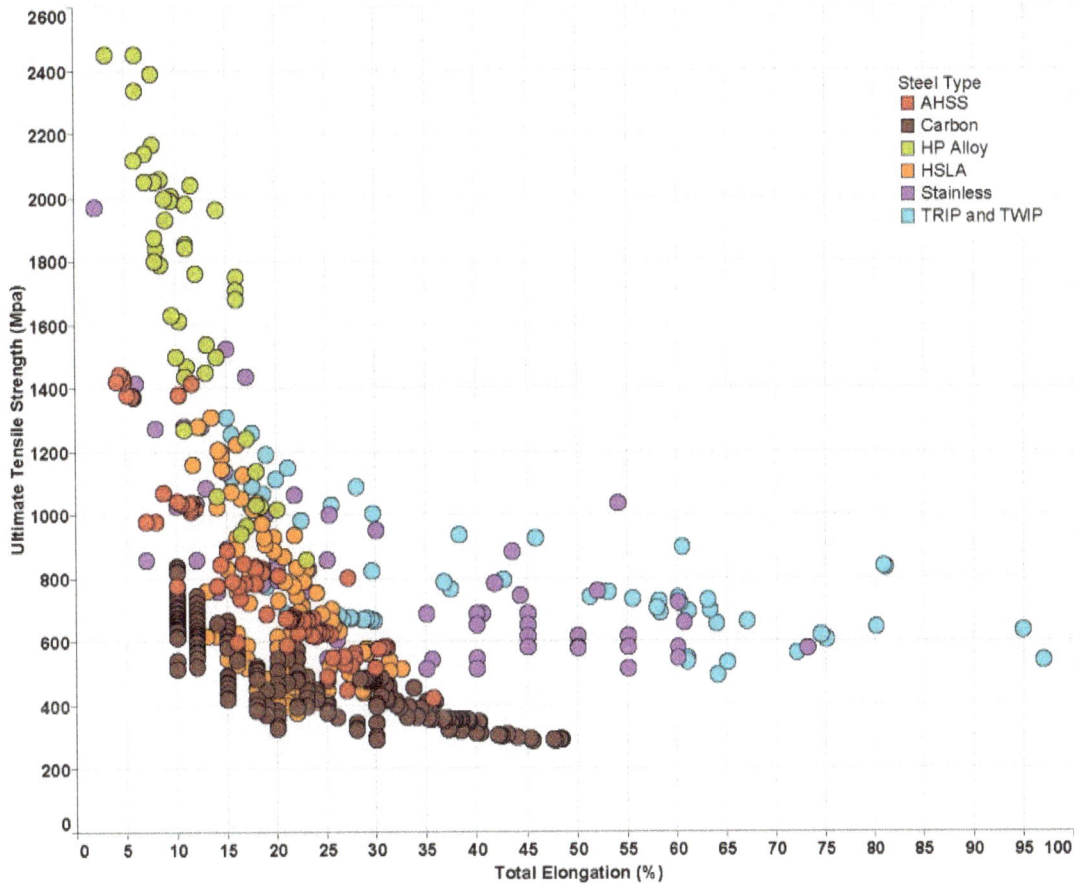

Almost all of the high strain rate data were compiled from published TSHB bar testing. However, there are a few instances where the high strain rate data were generated using a flywheel apparatus or compressive SHPB. These instances are noted throughout the report.

Figure 4 shows that tensile strength typically increases at higher strain rates. However, there are a few instances, especially with brittle steels, where tensile strength appears to decrease. The change in ductility with an increase in strain rate varies greatly with chemical composition and processing methods. However, certain regions of mechanical performance emerge for each category of steel.

Figure 4. High-rate and quasi-static strain-rate dependent properties for steel.

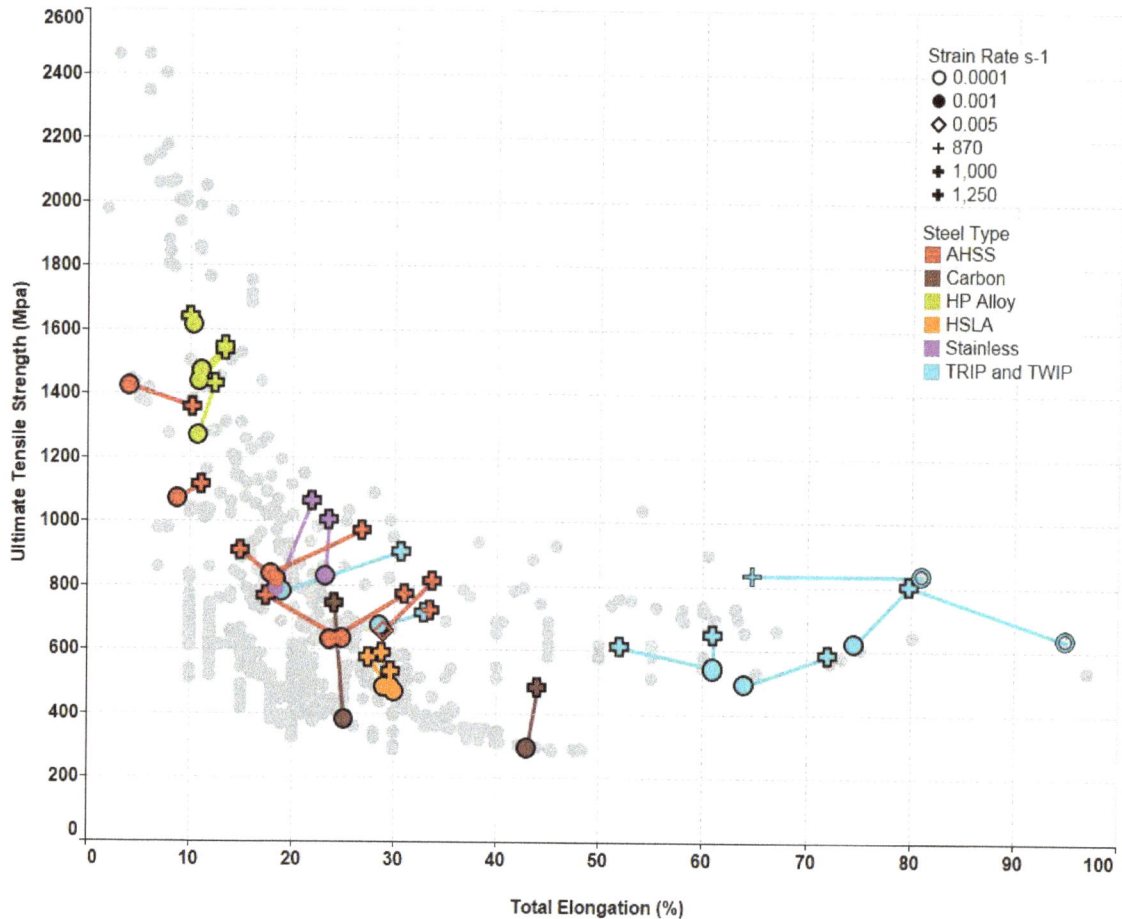

2.1.3   Microstructure

Steel microstructure is primarily dominated by three phases: austenite, martensite, and ferrite. Other phases may also be present depending on the chemical composition and processing method used to produce the steel. Each phase imparts different properties in the steel, and the presence of these phases has a profound effect on the strength, ductility, machinability, and durability of steel. The sections below briefly describe each of these phases.

2.1.3.1 Ferrite

α′, β, and δ ferritic phases are ferromagnetic phases that each have a body-centered cubic (bcc) crystalline structure. These phases are typically noted for having a very low carbon content of less than 0.01 % by weight (wt. %)[*]

---

[*] For a full list of the spelled-out forms of the units of measure used in this document, please refer to *U.S. Government Publishing Office Style Manual*, 31st ed. (Washington, DC: U.S Government Publishing Office, 2016), 248–252, https://www.govinfo.gov/content/pkg/GPO-STYLEMANUAL-2016/pdf/GPO-STYLEMANUAL-2016.pdf.

under ambient conditions. Excess carbon in the matrix readily forms a phase referred to as *cementite* (Cardarelli 2008).

2.1.3.2 Austenite

γ austenite is a non-ferromagnetic phase that typically has a face-centered cubic (fcc) crystalline microstructure. As carbon atoms accumulate, they occupy interstitial positions in the fcc microstructure, causing the lattice parameter to increase with an increase in carbon content (Cardarelli 2008).

2.1.3.3 Martensite

Martensite is an extremely hard and brittle phase that is vulnerable to corrosion-related deterioration. The amount of this phase that is formed is controlled by the carbon content of the steel and the cold-working methods that are used. In some steels, martensite and austenite-martensite phase transformations can occur as the steel is deformed. This bcc crystalline microstructure is ferromagnetic (Davis 1998).

2.1.3.4 Other phases

Cementite is a hard and brittle iron-carbide phase that forms during the cooling process in carbon-containing steels. The formation of cementite can be limited by increasing the rate of cooling (quenching) of the steel (Davis 1998).

2.2 Carbon steel

2.2.1 Overview

The term "carbon steel" is usually used to refer to either plain carbon steels or high-strength low-alloy (HSLA) carbon steels. Within each of these categories are steels that are produced in mass quantities for the construction and automotive industries.

Plain carbon steels are produced without the intentional addition of alloying elements other than carbon but have a lower combination of tensile strength and total elongation than most alloyed steels. HSLA steels are plain carbon steels that have select alloys added up to 5 wt. %. These alloys generally increase tensile strength and corrosion resistance of the steel. In many cases, these properties are improved by the addition of alloying elements without a reduction in ductility.

2.2.2   Plain carbon steel

2.2.2.1 Introduction

Plain carbon steels are generally more economical and easier to acquire locally than other steels. For example, ASTM A36 steel is a readily available and economical steel that is commonly used in the construction industry in the United States.

Figure 5 illustrates how the mechanical performance of plain carbon steel compares to other steel. In general, plain carbon steels have the lowest combination of strength and ductility.

Figure 5. Quasi-static ultimate tensile strength and total elongation properties for carbon steels.

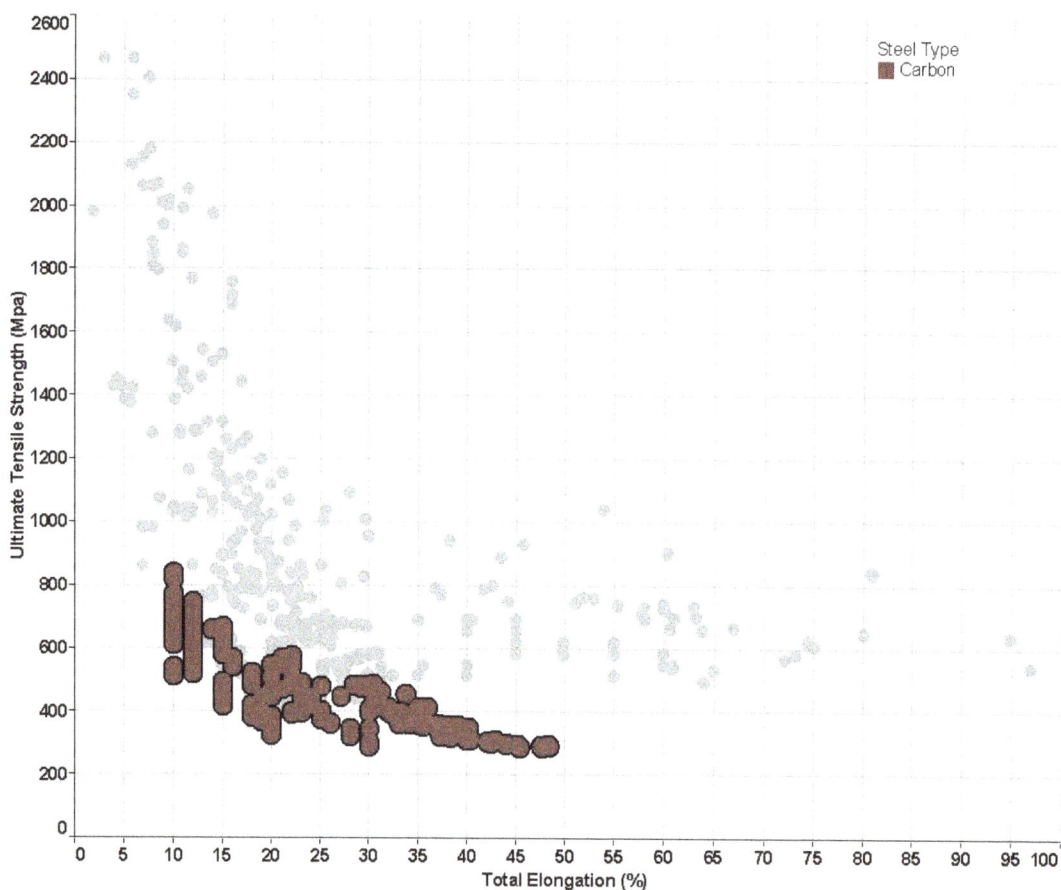

The data points used in Figure 5 were retrieved from Davis (1998), Harvey (1982), ASTM A 283, 285, 36, 529, 573, 575, 709, and 131, Sadagopan and Urban (2003), and the World Auto Steel website (Institute 2014).

2.2.2.2 AISI/SAE designation

AISI/SAE 10xx-plain carbon (0.00-0.99% Manganese)

-Low-carbon steels, also referred to as mild steels, contain less than 0.30 wt. % C. Comparatively, these steels have a lower ultimate tensile strength and greater ductility than other plain carbon steels. These steels include AISI-SAE grades 1005-1030 and ASTM A36 and are not considered to be hardenable through heat treatment. Due to their low carbon content, these steels are commonly cold-worked (Cardarelli 2008; Davis 1998).

-Medium-carbon steels contain between 0.31 and 0.55 wt. % C. These steels have a greater ultimate tensile strength than mild steels and lower ductility. Additionally, the strength and hardness of medium carbon steels can be increased by heat treatment. This group of steels includes AISI-SAE grades 1030 to 1055 (Cardarelli 2008).

-High-carbon steels theoretically contain between 0.56 and 1.00 wt. % C. However due to the difficulty of heat treating and cold processing high-carbon steels, carbon content is typically limited to less than 0.60%. These steels are generally brittle but may be further hardened through heat treatment. The hard, carbon-rich matrix typically provides good wear resistance (Davis 1998). This group of steels includes AISI-SAE grades 1055 to 1099.

AISI/SAE 11xx-resulfurized

See AISI/SAE 12xx section below.

AISI/SAE 12xx-resulfurized and rephosphorized

In some cases, phosphorus and sulfur may be reincorporated into the steel to improve certain properties. Phosphorus can increase tensile strength and corrosion resistance but can also greatly decrease ductility. Sulfur has little effect on longitudinal tensile strength but can make it more difficult to weld. Sulfur is usually reincorporated in machining steel. When machined, this type of steel crumbles away instead of curling or deforming (Davis 1998).

AISI/SAE 15xx-plain carbon (1.00-1.65% Manganese)

Manganese is common in steel, and manganese contents may be as high as 1.65 wt. % without intentionally adding manganese to the melt. The manganese content greatly impacts the strength and ductility of the steel, causing it to be divided into two categories. AISI-SAE 10xx steels have less than 1.00% manganese, while 15xx steels have 1.00-1.65% manganese.

2.2.2.3 Material Properties

Although the use of carbon steels in construction and manufacturing is prevalent, there has not been a lot of published research in this field during the last two decades. In fact, much of the research conducted on these materials exists in print only, making it more difficult to locate. Also, specifications are often written to describe only the minimum performance criteria for carbon steels. Published quality assurance testing typically only denotes if the steel met the criteria in the specification, not the actual (higher) ultimate tensile strengths and total elongation values. Therefore, the actual mechanical properties of carbon steels are more difficult to find in literature. Common material properties for plain carbon steels are in Table 2.

Table 2. Material properties of plain carbon steel (Cardarelli 2008).

Property	Value
Specific Gravity	~7.750
Elastic Modulus (GPa)	201-209
Shear Modulus (GPa)	81-82
Bulk Modulus (GPa)	160-170
Poisson's Ratio	0.27-0.30
Coefficient of Linear Thermal Expansion α: $\left(\frac{1}{K} \cdot 10^{-6}\right)$	11.1-12.8
Thermal Conductivity κ: $\left(\frac{W}{K \cdot m}\right)$	50.5-59.5
Electrical Resistivity ρ: $(\mu\Omega \cdot cm)$	14.2-17.0

2.2.2.4 Quasi-Static Tensile Properties

The effect that manganese content has on the mechanical properties of plain carbon steel is shown in Figure 6. The addition of manganese generally increases tensile strength and decreases ductility. The mechanical properties of carbon steels are also largely dependent on carbon content and processing methods. Carbon is important to the development of some strong microstructural phases such as martensite. Therefore, plain carbon steels with a high carbon content are generally higher in strength but more brittle. The effect of carbon content on ultimate tensile strength and ductility is depicted in Figure 7. Likewise, heat treatment, when combined with rapid quenching methods such as oil quenching, can improve tensile strength at the expense of ductility. However, it is worth noting that, for many applications, lower carbon ferritic steel is more formable and easier to produce (Cardarelli 2008, Davis 1998).

Figure 6. Effect of manganese content on mechanical properties of plain carbon steel.

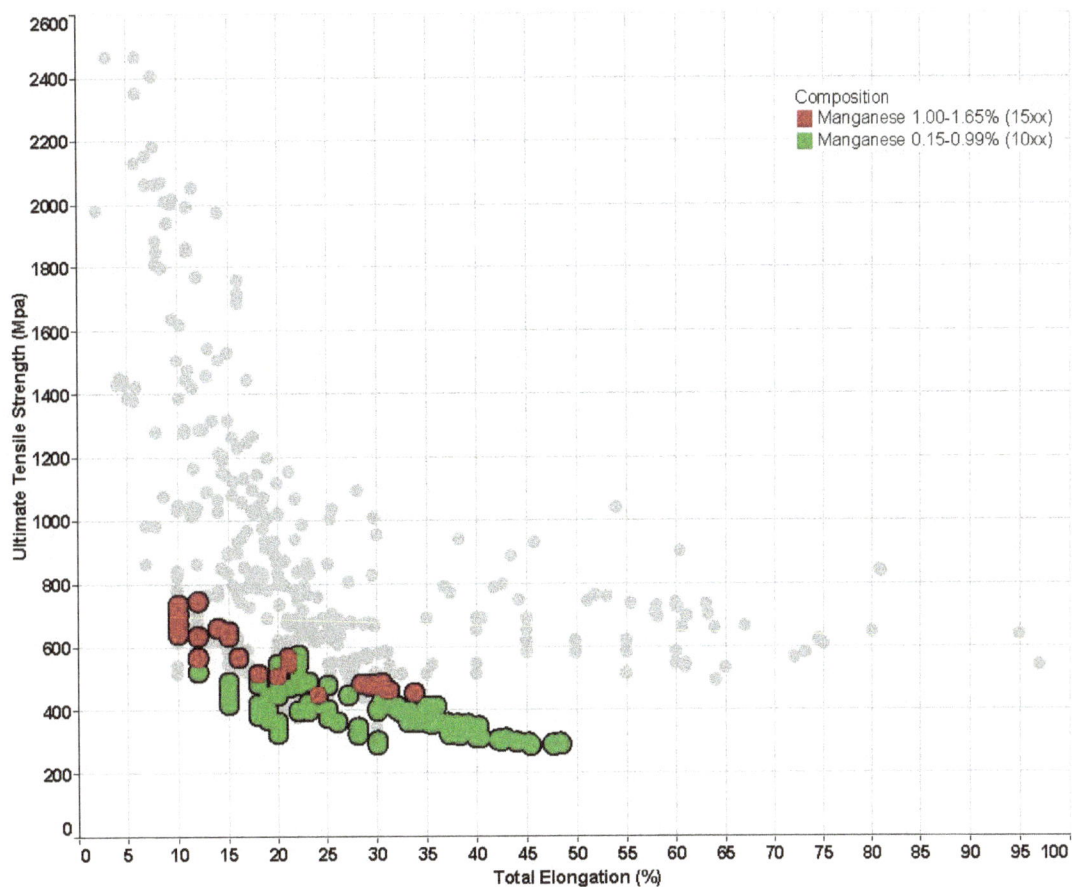

Figure 7. Effects of carbon content on the strength and ductility of plain carbon steels.

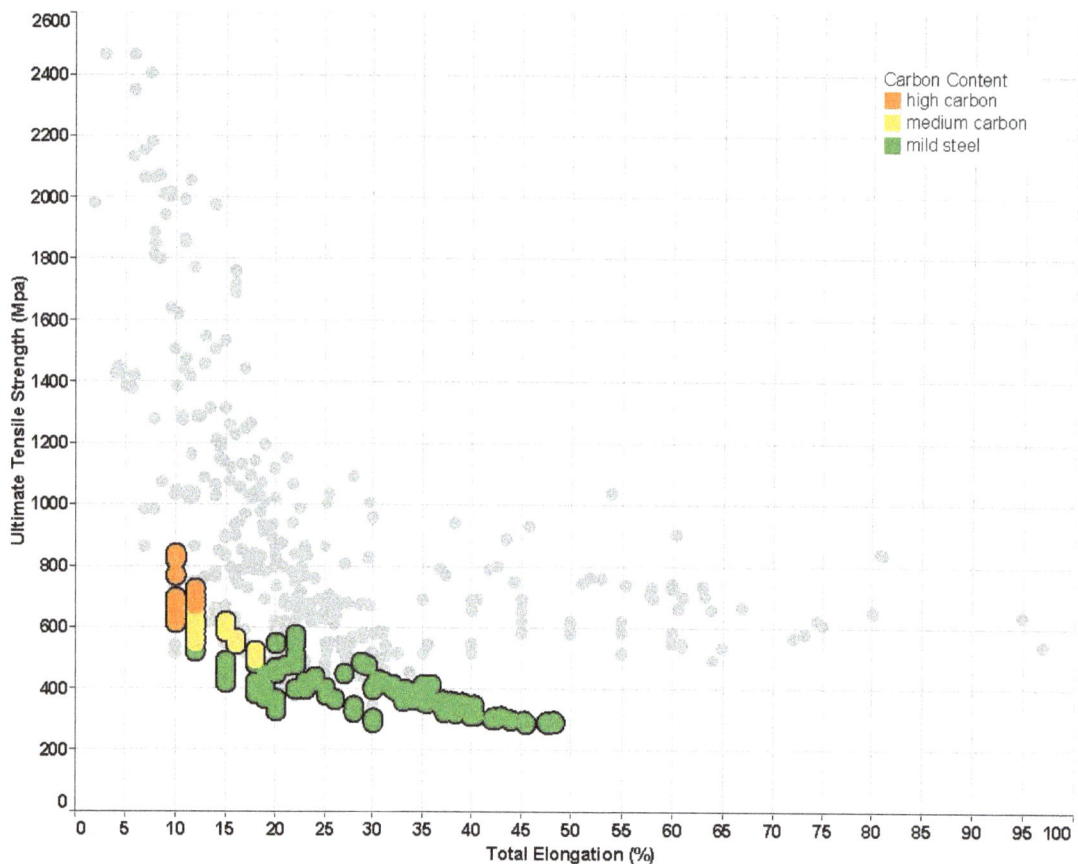

The data points used in Figure 6 and Figure 7 were retrieved from Davis (1998), Harvey (1992), ASTM A 283, 285, 36, 529, 573, 575, 709, and 131, Sadagopan and Urban (2003), and the World Auto Steel website (Institute 2014).

2.2.2.5 Dynamic tensile properties

The dynamic mechanical testing of carbon steel consistently yielded a higher ultimate tensile strength without a significant change in ductility. This behavior is shown in Figure 8.

Figure 8. Dynamic and quasi-static mechanical properties of plain carbon steel.

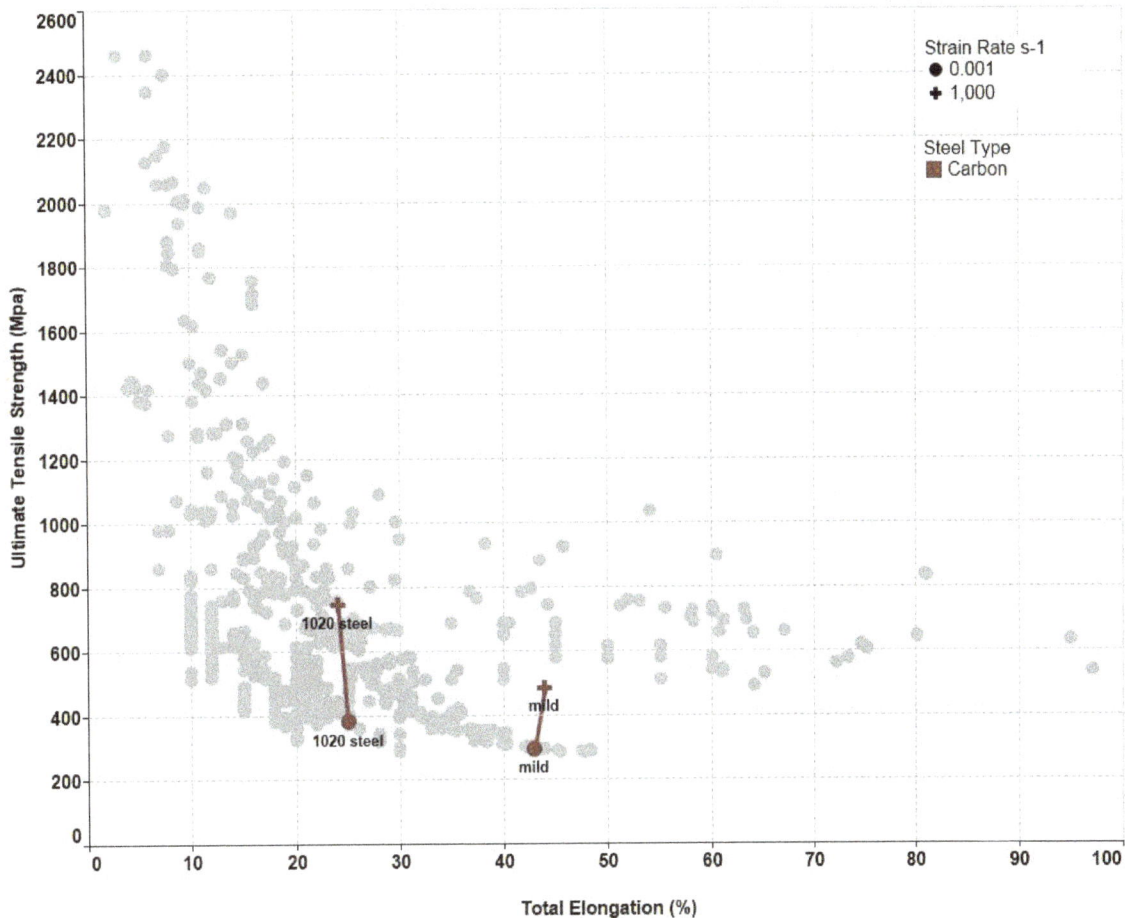

The data points used in Figure 8 were retrieved from the World Auto Steel website (Institute 2014).

2.2.3   Low-alloy and high-strength low-alloy carbon steels

2.2.3.1 Introduction

Low-alloy and high-strength low-alloy (HSLA) carbon steels have incorporated small amounts of other elements to improve mechanical, corrosion, or formability properties. These alloys typically make up less than 5% of the material weight but, as is shown in Figure 9, the effect that these alloys can have on mechanical properties is significant. The band of mechanical performance depicted in Figure 9 has a significantly higher ultimate tensile strength than similar plain carbon steels that were shown in Figure 5.

Figure 9. Quasi-static ultimate tensile strength and total elongation properties for low-alloy steels.

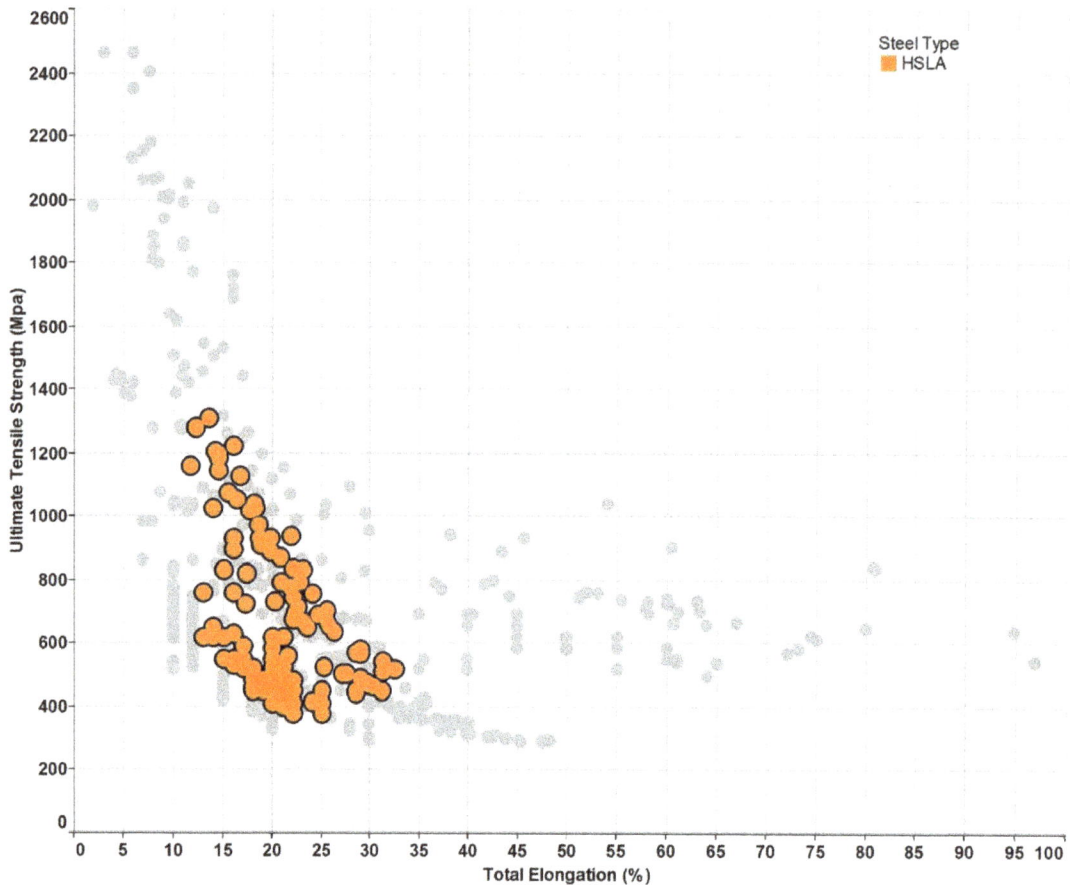

The data points used in Figure 9 were retrieved from Davis (1998), Harvey (1992), ASTM A 283, 285, 36, 529, 573, 575, 709, and 131, Sadagopan and Urban (2003), Society of Automated Engineers (2008), and the World Auto Steel website (Institute 2014).

2.2.3.2 Effect of Alloy Elements [Cardarelli 2008, Davis 1998, Sadagopan and Urban 2003]

As mentioned in the plain carbon section, manganese is found in most steels. During production it helps to deoxidize the melt and reduce susceptibility to hot shortness, a quality that causes the steel to crack or tear during hot rolling or welding. When combined with sulfur in the right proportions, manganese improves the machinability of steel. Manganese also slightly increases the tensile strength of steel.

Silicon is used to help deoxidize steel. It can increase the strength of ferrite phases with only a minimal loss of ductility. In exceptionally low-carbon and low-carbon resulfurized alloys, the addition of silicon can result in surface defects.

Chromium increases corrosion and oxidation resistance. Chromium is considered a hardening alloy that produces strong chromium-iron-carbide phases that help improve strength at high temperatures and provide abrasion resistance.

Nickel is often used in conjunction with chromium in high-strength and high-performance steels. Nickel strengthens ferrite and does not form carbide compounds in steel. Nickel also improves low-temperature strength and toughness.

Molybdenum reduces the susceptibility of steel to temper embrittlement. It works to slow the transformation of austenite to pearlite and increases tensile strength at high temperatures.

Vanadium increases tensile strength by forming strong carbide phases.

Tungsten is frequently used in high-speed tool steels because steels alloyed with tungsten are heat-resistant. Tungsten carbide prevents softening during the tempering process.

2.2.3.3 Trace alloy additions [Cardarelli 2008, Davis 1998, Sadagopan and Urban 2003]

Aluminum is commonly used to deoxidize the steel melt and control the grain size of austenite.

Boron (0.0005-0.003 wt. %) helps improve hardenability and corrosion resistance.

Zirconium limits grain growth and helps deoxidize steel.

Lead (0.15 to 0.35 wt. %) is added to improve machining properties.

Nitrogen is typically incorporated during the melting process to reduce the potential for oxidation. Nitrogen can form aluminum nitride that helps control grain size. Nitrogen can negate some of the benefit achieved from incorporating boron alloys.

Titanium is often used in conjunction with boron because it combines with oxygen and nitrogen in the steel, both of which reduce the effectiveness of the boron alloy. Titanium also helps to deoxidize the melt and limit grain growth.

Calcium helps to deoxidize steels and control the shape of nonmetallic inclusions. This results in a tougher steel that exhibits improved machinability.

2.2.3.4 AISI/SAE/ASTM designations

AISI/SAE/ASTM designations for low alloy and HSLA steels are listed in Table 3.

Table 3. Low alloy and HSLA steel designations.

Specification (SAE/AISI/ASTM)	Criteria
13xx	>1.75% Mn
23xx	~3.50% Ni
25xx	~5.00% Ni
31xx	~1.25% Ni; 0.65-0.80% Cr
32xx	~1.75% Ni; ~1.07% Cr
33xx	~3.50% Ni; 1.50-1.57% Cr
34xx	~3.00% Ni; ~0.77% Cr
40xx	0.20-0.25% Mo
44xx	0.40-0.52% Mo
41xx	0.50-0.95% Cr; 0.12-0.30% Mo
43xx	~1.82% Ni; 0.50-0.80% Cr; 0.12-0.25% Mo
47xx	~1.05% Ni; ~0.45% Cr; 0.20-0.35% Mo

Specification (SAE/AISI/ASTM)	Criteria
81xx	~0.30% Ni; ~0.40% Cr; ~0.12% Mo
86xx	~0.55% Ni; ~0.50% Cr; ~0.20% Mo
87xx	~0.55% Ni; ~0.50% Cr; ~0.25% Mo
88xx	~0.55% Ni; ~0.50% Cr; ~0.35% Mo
93xx	~3.25% Ni; ~1.20% Cr; ~0.12% Mo
94xx	~0.45% Ni; ~0.40% Cr; ~0.12% Mo
97xx	~0.55% Ni; ~0.20% Cr; ~0.20% Mo
98xx	~1.00% Ni; ~0.80% Cr; ~0.25% Mo
46xx	0.85-1.82% Ni; 0.20-0.25% Mo
48xx	~3.50% Ni; ~.25% Mo
50xx*	0.27-0.65% Cr
51xx*	0.80-1.05% Cr
61xx	0.60-0.95% Cr; 0.10-0.20% V
72xx	~1.75% W; ~0.75% Cr
92xx	1.40-2.00% Si; 0.65-0.85% Mn; 0.00-0.65% Cr
9xx	High Strength Low Alloy (various SAE grades)
ASTM Designations:	A242, A572, A588, A606, A607 (withdrawn 2000), A618, A633, A656, A690, A847, A871, A913, A992, A514, A709, A710, A1077

*A minimum of 1.00% carbon is required by AISI for bearing steels.

When high-strength low-alloy (HSLA) steels are specified by ASTM, they may be specified by either mechanical properties or chemical composition. HSLA steels are low in carbon and typically have an alloy content less than 2 wt. %. (Davis 2001).

2.2.3.5 Material properties

Materials properties for alloy-carbon steels are in Table 4.

Table 4. Material properties of alloy-carbon steel (Cardarelli 2008).

Property	Value
Specific Gravity	7.750
Elastic Modulus (GPa)	201-209
Shear Modulus (GPa)	81-82
Bulk Modulus (GPa)	160-170
Poisson's Ratio	0.27-0.30
Coefficient of linear Thermal Expansion α: $\left(\frac{1}{K} \cdot 10^{-6}\right)$	10.9-12.8
Thermal Conductivity κ: $\left(\frac{W}{K \cdot m}\right)$	34.0-48.6
Electrical Resistivity ρ: $(\mu\Omega \cdot cm)$	22.0-30.0

2.2.3.6 Quasi-static tensile properties

Processing methods and carbon content have a particularly large influence on the mechanical properties of low-alloy steels. This is demonstrated in Figure 9 by the band of steels referred to as "Alloy Steels" just exceeding the mechanical performance of plain steel. However, most of the steels occupy the same space regardless of alloying element. This behavior is shown in Figure 10.

Many of the ASTM standards are written as minimum requirements for this group of steels; therefore, these values are expected to be below the actual experimental results. In Figure 10, the ASTM and SAE standards are lower in strength and ductility than the experimental data.

Figure 10. Effects of alloy additions on low-alloy and high-strength low-alloy steels.

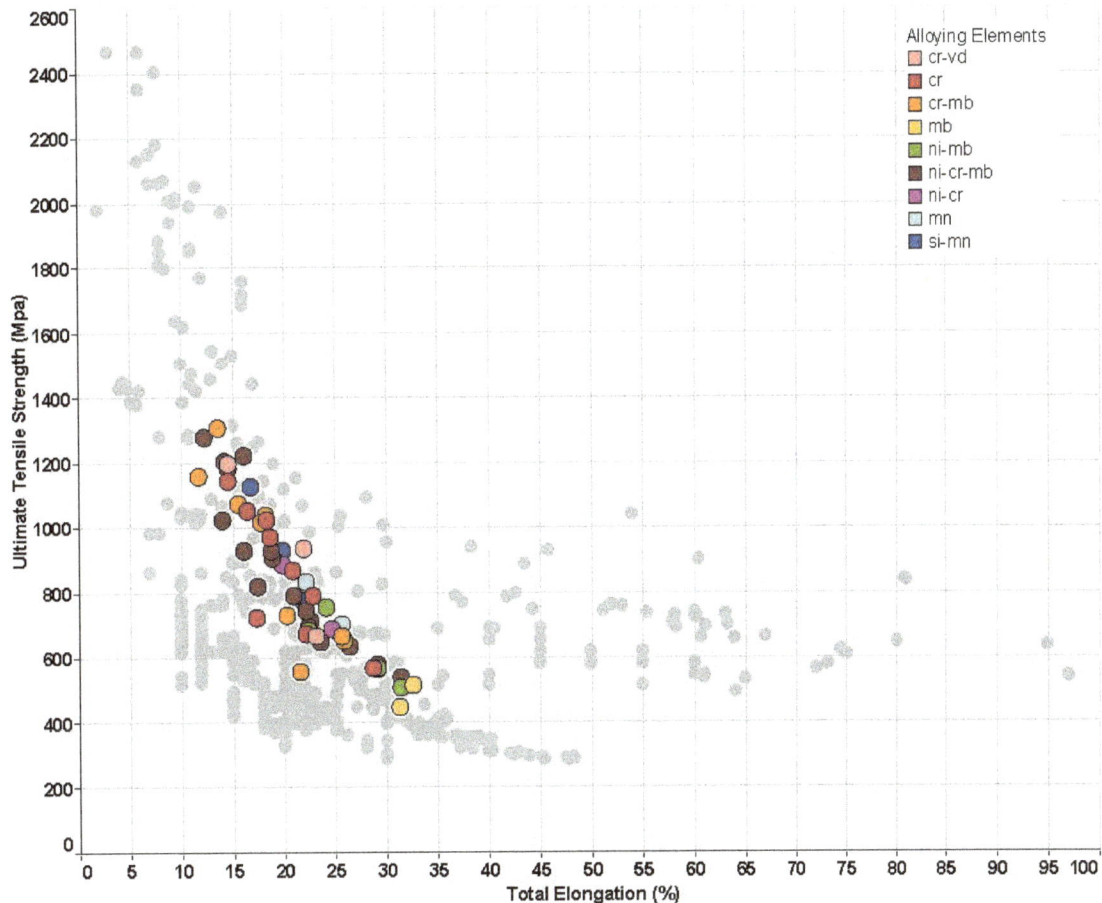

The data points used in Figure 10 were retrieved from Davis (1998), Harvey (1972), ASTM A 514, 710, 572, 242, 618, 913, 992, 709, 588, 606, 633, 656, 847, and 871, Sadagopan and Urban (2003), Society of Automated Engineers (2008), and the World Auto Steel website (Institute 2014).

2.2.3.7 Dynamic tensile properties

Only limited data are available on the mechanical performance of low-alloy and HSLA steels tested under dynamic conditions. Figure 11 shows that the high-strength, low-alloy steel with a yield strength of 350 MPa was only moderately strain-rate sensitive; and, like plain carbon steels, the HSLA 350 steel did not experience a great change in ductility from quasi-static to dynamic strain rates.

Figure 11. Dynamic and quasi-static mechanical properties of low-alloy steel.

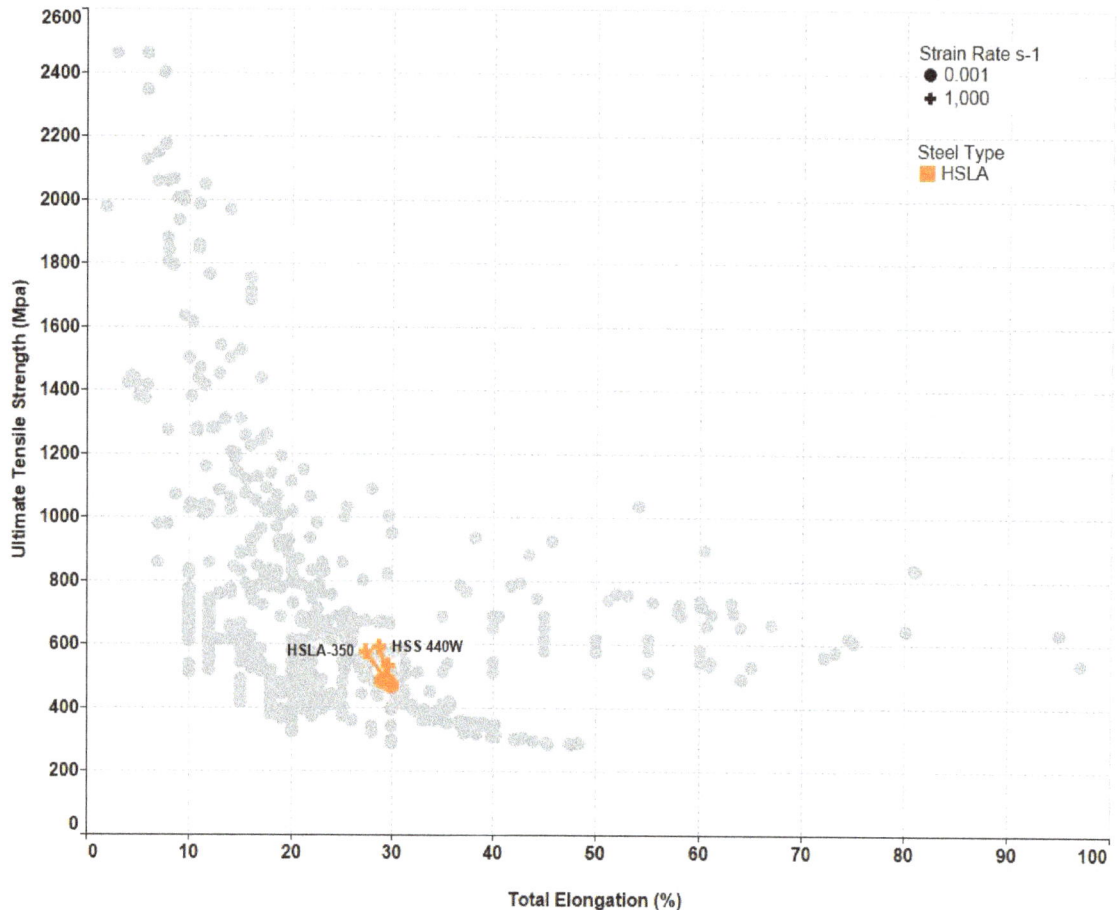

The data points used in Figure 11 were retrieved from the World Auto Steel website (Institute 2014).

2.3 Advanced high-strength steels

Advanced high-strength steels (AHSS) are being used more frequently in the automotive industry. Most AHSS are higher in tensile strength and comparable in ductility to plain carbon steels. However, there are some AHSS that are very ductile. This high ductility is attributed to a high manganese alloy content that promotes mechanisms described as twinning-induced plasticity (TWIP) and transformation-induced plasticity (TRIP).

AHSS can withstand more deformation and, in most instances, will absorb more energy than plain carbon steel. For this reason, many of these steels are used by the automotive industry to improve crashworthiness (Sadagopan and Urban 2003; W. Wang et al. 2013).

AHSS has a chemical composition similar to low-carbon steel except that it has been alloyed with 2-40 wt. % manganese. Adding manganese reduces the overall weight of the steel. Steels in this category are also often alloyed with silicon and aluminum, and they are identified by a combination of the phase(s) present, minimum tensile strength, and dislocation mechanics.

In this report, stainless steels are separated from AHSS due to differences in chemical composition and corrosion resistance. However, similar mechanical properties are observed in both steel categories. AHSS are generally alloyed with manganese while stainless steels are alloyed with chromium.

2.3.1   Martensitic, ferritic, dual phase, and complex phase

2.3.1.1 Introduction

Steels that are classified as martensitic (M), ferritic (F), dual phase (DP), or complex phase (CP) and were alloyed with manganese fall into this steel category. These steels are higher in tensile strength than plain carbon steels but comparable in ductility, as evidenced by the mechanical properties shown in Figure 12.

Most literature focuses on the greater tensile strength of martensitic and ferritic steels. Austenitic steels are typically only discussed in high manganese TWIP steels. M-, F-, DP-, and CP-phase steels were separated from TRIP and TWIP steels in this report due to differences in deformation mechanics.

Figure 12. Quasi-static ultimate tensile strength and total elongation properties for advanced high-strength steels.

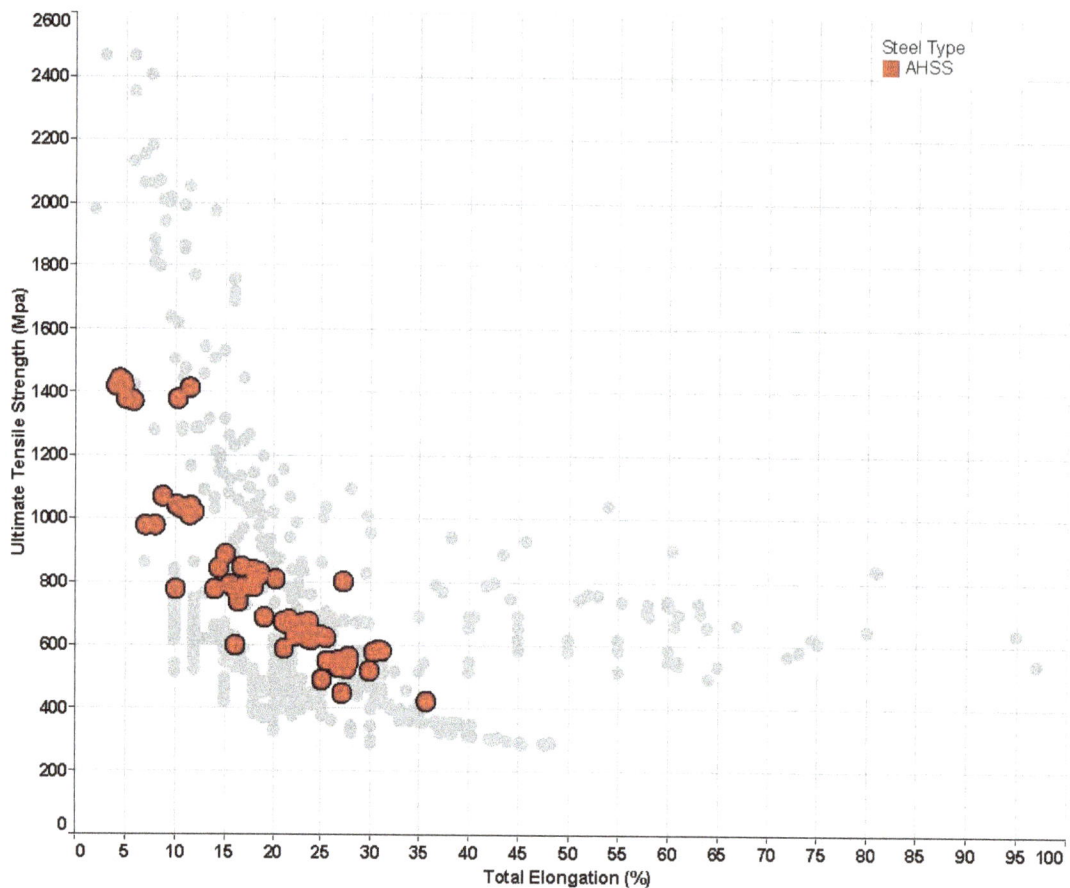

The data points used in Figure 12 were retrieved from Sadagopan and Urban (2003), the World Auto Steel website (Institute 2014), Tavares et al. (1999), W. Wang et al. (2013), ASTM A1088, and the ASM Handbook Vol. 4 (1991).

2.3.1.2 Designations

Martensitic (M) steels have a high yield strength and ultimate tensile strength but poor ductility and corrosion resistance. Martensitic steels also tend to crack when welded (Cardarelli 2008; Davis 1998; and Boh et al. 2004).

Dual-phase (DP) steels are generally a combination of martensite and ferrite phases and feature a curved stress strain curve similar to other specialized steels (Cardarelli 2008). Unlike duplex stainless steels, dual-phase AHSS contain little or no chromium; therefore, these steels do not exhibit the high corrosion resistance of stainless steel.

Complex Phase (CP) steels, like DP steels, contain little or no chromium. These steels possess a combination of martensitic, ferritic, and austenitic phases. CP steels have similar (or slightly improved) corrosion resistance when compared to DP steels; however, corrosion resistance is still poor.

2.3.1.3 Quasi-static tensile properties

Keeler (2014) uses Figure 13 to demonstrate the mechanical performance of a dual-phase steel compared to a conventional high-strength low-alloy (HSLA) steel. Both steels have the same nominal yield stress of 350 MPa and similar elongation to failure. However, the DP steel has an ultimate tensile strength approximately 100 MPa higher than the HSLA steel.

Figure 13. UTS and elongation to failure of DP 350/600 and HSLA 350/450 under quasi-static loading conditions (.01s^{-1}) from Keeler (2014).

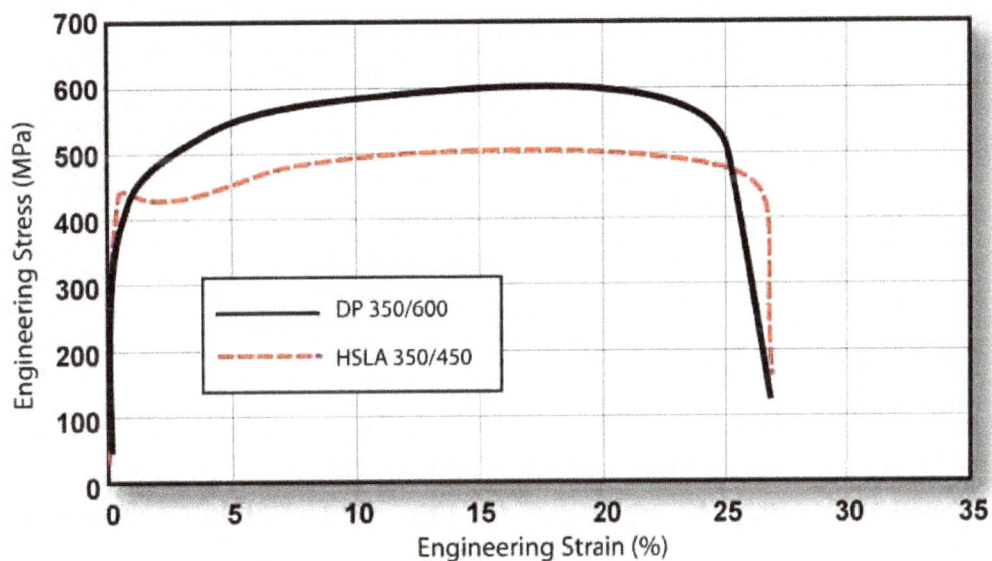

The rounded shape of the DP curve is also common in TRIP and TWIP steels. This observation has led many authors to question the applicability of the strain-hardening exponent "n", commonly used in the Holloman equation, i.e., $\sigma = K\varepsilon^n$ (Sadagopan and Urban 2003; Keeler 2014; Billur and Altan 2012).

The trends in Figure 14 indicate that steels higher in martensite are generally higher in tensile strength and more brittle. This behavior is expected due to a combination of the phases present, high carbon content, and processing methods.

Figure 14. Common advanced high-strength steel designations.

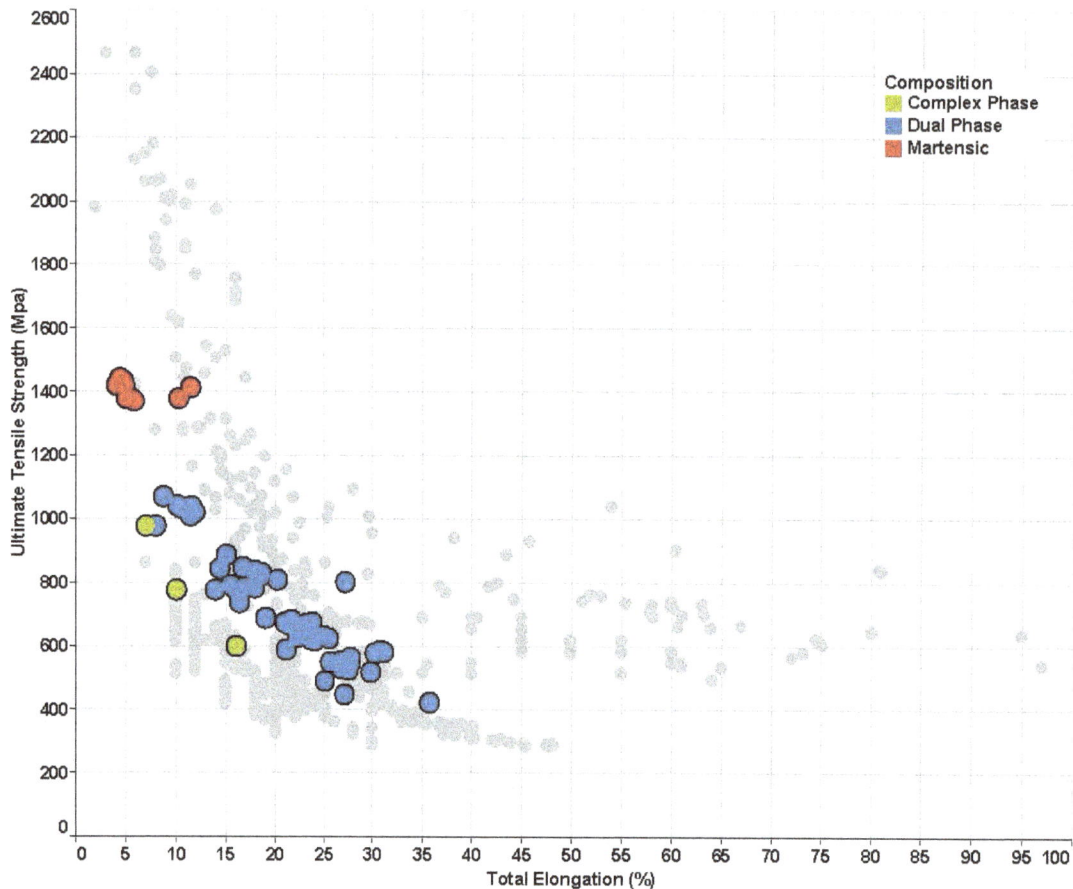

The data in Figure 14 were retrieved from Sadagopan and Urban (2003), the World Auto Steel website (Institute 2014), Tavares et al. 1999, W. Wang et al. 2013, ASTM A1088, and the ASM Handbook Vol. 4 (1991).

2.3.1.4 Dynamic tensile properties

From the data compiled into Figure 15, steels containing greater amounts of martensite appear to have tensile strengths that were less strain-rate dependent than more ferritic steels. The steels with high martensite contents also appear to experience a slight increase in total elongation from quasi-static to dynamic strain rates. This behavior is likely a function of the brittle failure that these metals experience under quasi-static conditions.

The data on more ductile, DP steels came from multiple reports. The authors agree that the ultimate tensile strength of the DP steels increased at higher strain rates. However, some authors observed that these steels were more brittle at high strain-rates (W. Wang et al. [2013]

and Wong [2005]), while others observed more ductile behavior (Wong [2005] and Institute [2014]).

Figure 15. Dynamic and quasi-static mechanical properties of advanced high-strength steels.

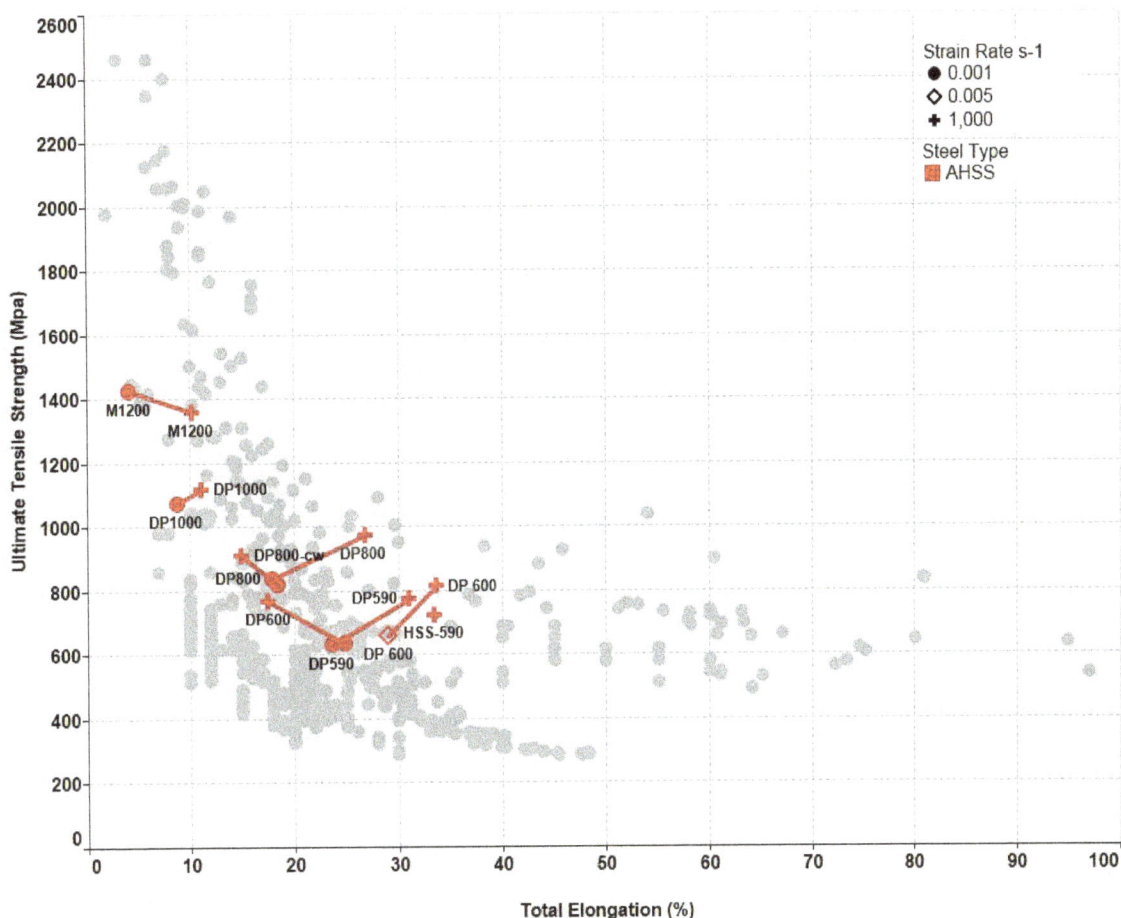

The data points used in Figure 15 were retrieved from W. Wang et al. (2013), the World Auto Steel website (Institute 2014), and Wong (2005).

Wong (2005) found that, when compared to quasi-static loading conditions, the elongation to failure is much greater under dynamic conditions (1000 s^{-1}) for dual phase and martensitic steels where the primary phase was martensite. Also, the ultimate tensile strength of the primarily martensitic steels remained constant despite changes in strain rate (.001 to 1250 s^{-1}). The stress-strain plot reported by W. Wang et al. (2013) for a M1200 martensitic steel is shown in Figure 16.

Figure 16. Tensile engineering stress-engineering strain plot of a M1200 steel tested at various strain rates (from W. Wang et al. 2013).

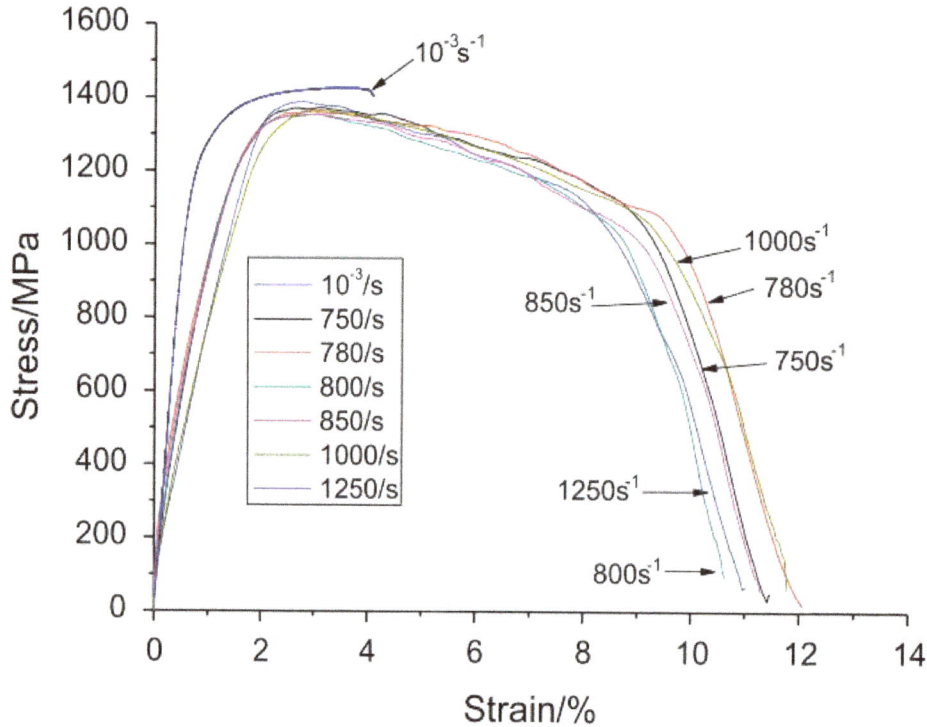

Additionally, W. Wang et al. (2013) found that, for dual phase steels where the primary phase was ferrite, the elongation to failure was reduced under dynamic conditions (1,000 s^{-1}) when compared to quasi-static loading conditions. However, the ultimate tensile strength of these steels still increased by 15-20% under dynamic loading conditions. The results of a DP600 steel that was tested under both quasi-static and dynamic TSHB loading conditions are shown in Figure 17.

Wong published the results of an International Iron and Steel Institute (IISI) Round-Robin laboratory testing program that included high strain-rate testing of both mild steel and DP 590 steel. Of the three laboratories that achieved strain rates near 1,000 s^{-1}, one used a servo-hydraulic testing machine, one used a TSHB, and the other used a single bar apparatus. The TSHB results are included in the data in Figure 15. The results achieved by the laboratory using the servo-hydraulic apparatus were comparable in ultimate tensile strength and had greater elongation to failure than the laboratory using the TSHB. The servo-hydraulic system also recorded a significant amount of vibration. The laboratory that used the single bar test apparatus reported significantly lower ultimate tensile strength and total elongation results Wong (2005). The data set that was published by Wong is a good example of why it is important to understand the procedures used in the setup, testing, and data analysis of high strain-rate tests. A plot of high strain rate results for the DP590 steel from this report is shown in Figure 18.

Figure 17. Tensile engineering stress-engineering strain plot of a DP600 steel tested at various strain rates (from W. Wang et al. 2013).

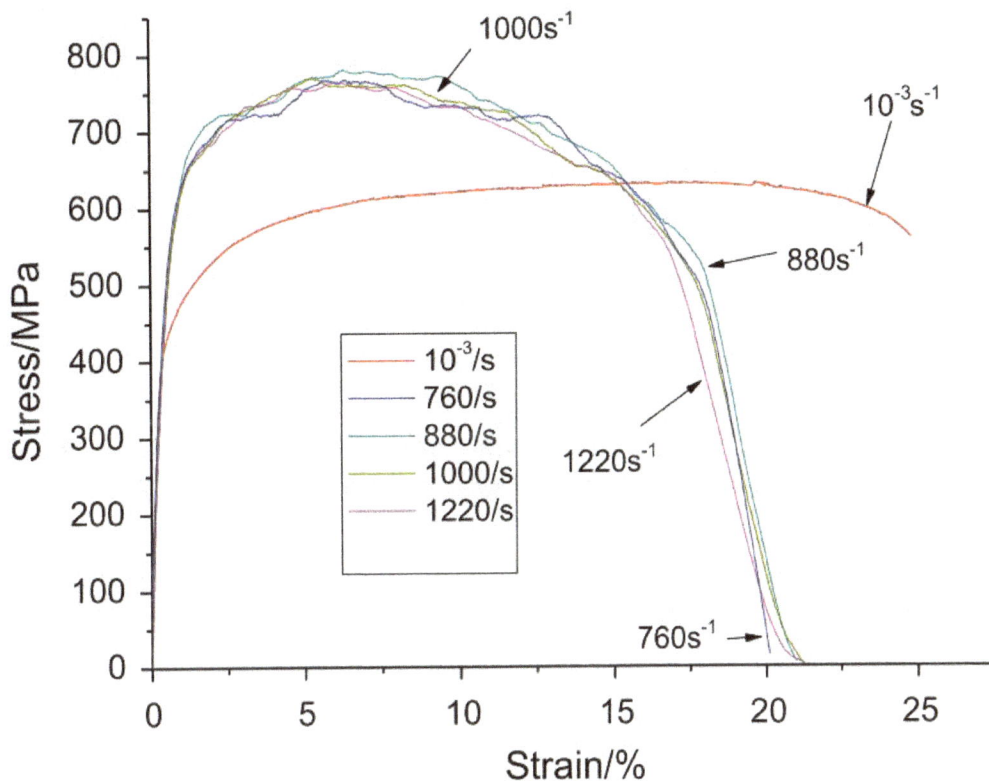

Huh et al. (2008) reported that, for DP800 steel, the elongation to failure remained constant and, for DP600 steel, the elongation to failure increased slightly (6%) when subject to tensile loads applied at strain rates ranging from 0.003 to 200 s^{-1}.

Figure 18. Tensile engineering stress-engineering strain plot of a DP590 steel tested at various strain rates (from Wong 2005).

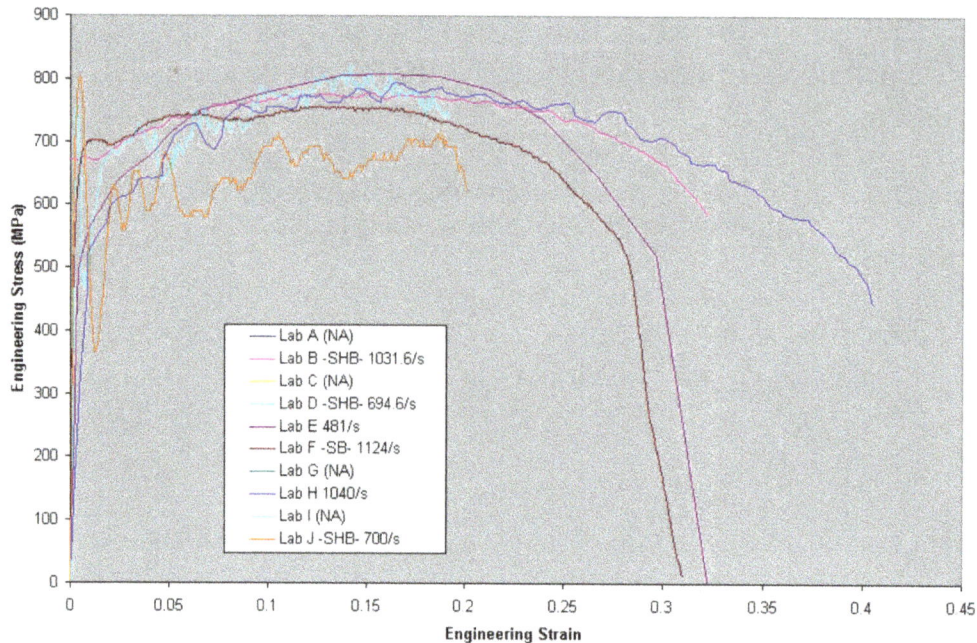

Data obtained from the World Auto Steel website (Institute 2014) showed that ductility improved at high strain rates for both DP600 and DP800 steels.

Kim et al. (2013) studied the applicability of the Holloman and Voce hardening equations for the stress-strain behavior of DP 780 and DP 980 steels at rates from .001 to 200 s^{-1}.

2.3.2 TRIP and TWIP steels

2.3.2.1 Introduction

Twinning-induced plasticity (TWIP) and transformation-induced plasticity (TRIP) are designations that indicate a phase transformation within the microstructure has an impact on the mechanical properties of the steel. TWIP steels are primarily austenitic steels that form mechanical twins as the material experiences stress-induced deformation. The twinning mechanism results in an extremely ductile metal with moderate-to-high tensile strength.

However, research has shown that processing methods such as cold rolling can reorient the microstructure of some of these steels, making it more difficult for twins to form (Dini et al. 2010). During work hardening, some of the austenitic phases are transformed from (γ) austenite to (α′)

martensite. An intermediate phase (ε) martensite may also form. Such phase transformations also occur during mechanical testing, resulting in the term TRIP steel. Steels in this category may undergo either or both of these types of deformations (De et al. 2006; Milad et al. 2008).

Similar deformation mechanisms are also sometimes seen in stainless steels. However, TRIP and TWIP steels are generally high in manganese (15-40 wt. %), where stainless steels are instead alloyed with chromium (12-35 wt. %). The addition of aluminum suppresses TRIP and TWIP mechanisms whereas the addition of silica reduces the stacking fault energy and promotes austenite-martensite transformation (Milad et al. 2008; Frommeyer et al. 2003). Figure 19 shows the quasi-static strength and elongation data for TRIP and TWIP steels.

Figure 19. Quasi-static ultimate tensile strength and total elongation properties for TRIP and TWIP steels.

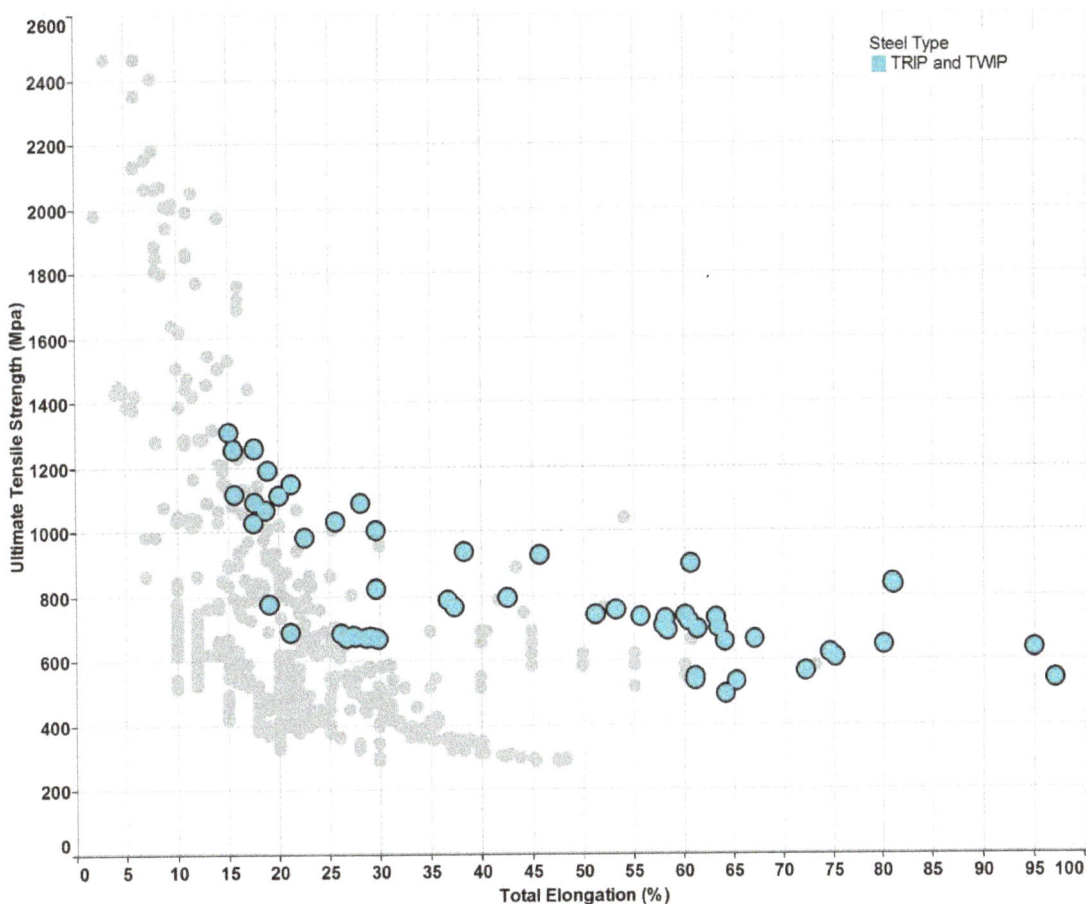

The data points shown in Figure 19 were retrieved from Frommeyer et al. (2003), Grässel et al. (2000), ASTM A 1088, Ding et al. (2006),

Sadagopan and Urban (2003), the World Auto Steel website (Institute 2014), Curtze and Kuokkala (2010), and Dini et al. (2010).

2.3.2.2 Classification

In specifications, TRIP and TWIP steels are typically designated by the dominant deformation mechanism (either TRIP or TWIP) and by either the yield strength or ultimate tensile strength.

For example, ASTM A 1088 specifies TRIP 780T/500Y where the primary deformation mechanism is transformation-induced plasticity, its minimum ultimate tensile strength is 780 MPa, and its minimum yield strength is 500 MPa.

However, many steels undergo both transformation- and twinning-induced plasticity, and the extent of each type of deformation is a function of both chemical composition and processing method. Therefore, many researchers reference these steels by manganese alloy content and add a note about the processing method(s) used.

2.3.2.3 Quasi-static tensile properties

TRIP and TWIP steels vary considerably in ductility. The effect that the manganese-alloy content and annealing temperature have on the tensile strength and ductility of TRIP and TWIP steels is shown in Figure 20 and Figure 21. In TRIP and TWIP steels, tensile strength decreases slightly with an increase in manganese alloy content, but ductility increases considerably with increases in manganese alloy content and annealing temperature. This is different from carbon steels where the manganese content is low and the addition of manganese, even in small amounts, resulted in an increase in strength.

Curtze and Kuokkala (2010) also found that niobium is capable of increasing tensile strength without reducing ductility in TWIP steels.

Ductility decreases when the steel is tested at elevated temperatures. In fact, several authors found that the elongation to failure under quasi-static conditions was maximized for austenitic TWIP steels when the temperature of the metal at the time of loading was between -50°C and 0°C (Frommeyer, Brüx, and Neumann 2003, Grässel et al. 2000, Curtze and Kuokkala 2010).

In addition to studying the effects of annealing temperature, Dini et al. (2010) also studied the effect of cold-working a 31 wt. % Mn AHS steel. These results are summarized in Figure 21. These data help quantify the degree to which cold reduction increases tensile strength and reduces ductility in high Mn alloy steels.

Figure 20. TRIP and TWIP steel designations.

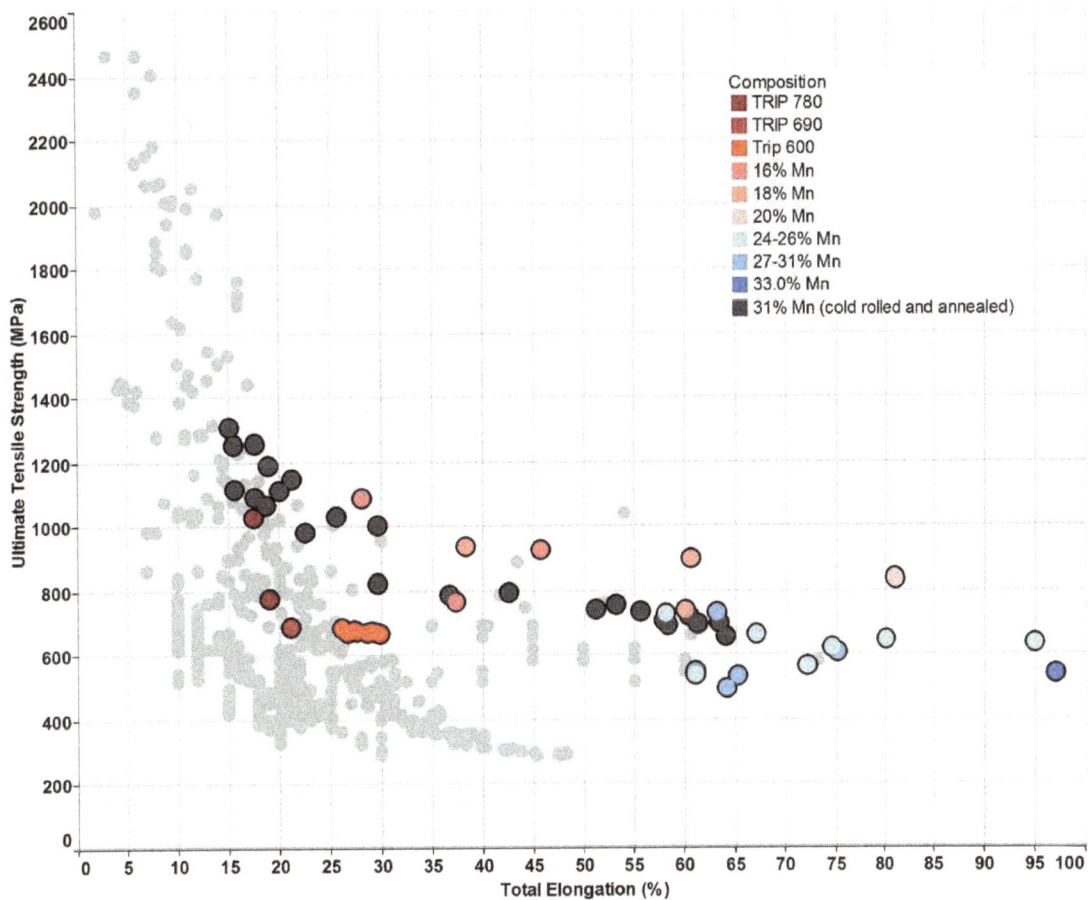

The data in Figure 20 were obtained from Frommeyer et al. (2003), Grässel et al. (2000), ASTM 1088, Ding et al. (2006), Sadagopan and Urban (2003), the World Auto Steel website (Institute 2014), Curtze and Kuokkala (2010), and Dini et al. (2010).

Figure 21. Treatment of a 31%-Mn TWIP steel.

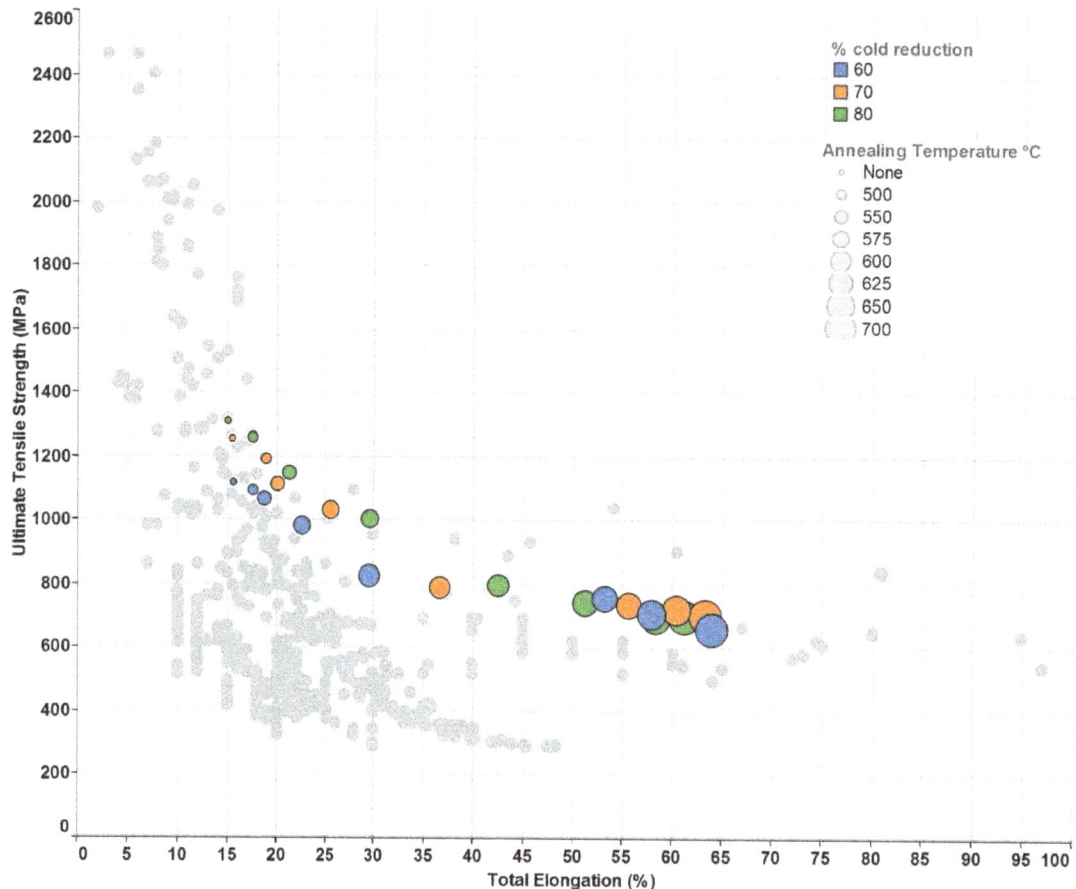

The data used in Figure 21 were obtained from Dini et al. (2010).

Frommeyer et al. (2003) compared the mechanical performance of 15.8% Mn (TRIP), 20.1% Mn (TRIP/TWIP), and 25.6% Mn (TWIP) steels. In that article, the purely austenitic 25.6% Mn TWIP steel remained purely austenitic after tested to failure as determined through x-ray diffraction. A large percentage of the matrix of the remaining two steels was transformed from the ε-martensite and austenite phases to α'-martensite. This transformation resulted in a high-strength material with moderate elongation to failure and distinct strain-hardening properties. This behavior is evidenced by the steep increase in true stress reported by Frommeyer et al. in Figure 22.

Figure 22. True stress-strain curves from Frommeyer et al. (2003).

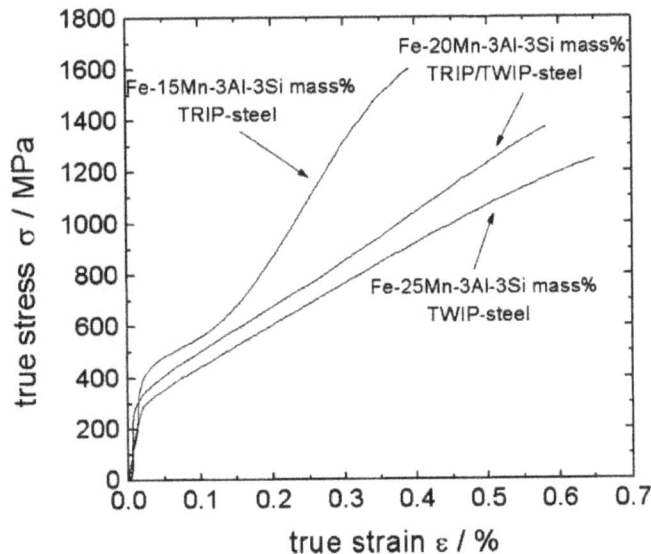

Grässel et al. (2000) noted similar strain-rate behavior in TWIP steels under quasi-static to intermediate strain rate conditions as Das et al. noted for a 304LN stainless steel. Test results for the stainless steel from Das et al. (2008) are shown in Figure 30 in the stainless steel section. This behavior shows a significant reduction in elongation to failure for strain rates ranging between .0001 and 1 s^{-1} (Grässel et al. 2000; Das et al. 2011). Grässel et al. attributed similar behavior in TWIP steel to adiabatic heating of the specimen during tensile testing.

2.3.2.4 Dynamic tensile properties

A summary of the dynamic ultimate tensile strength and total elongation properties of TWIP and TRIP steels obtained from published literature are shown in Figure 23. These data were obtained from conventional TSHB tests, torsional SHB tests, and fly-wheel tests. The test results from the TSHB show that, at high strain rates, TRIP and TWIP steels increase in strength and maintain or slightly increase in ductility when compared to their response under quasi-static loading conditions.

Figure 23. Dynamic and quasi-static mechanical properties of TWIP steel.

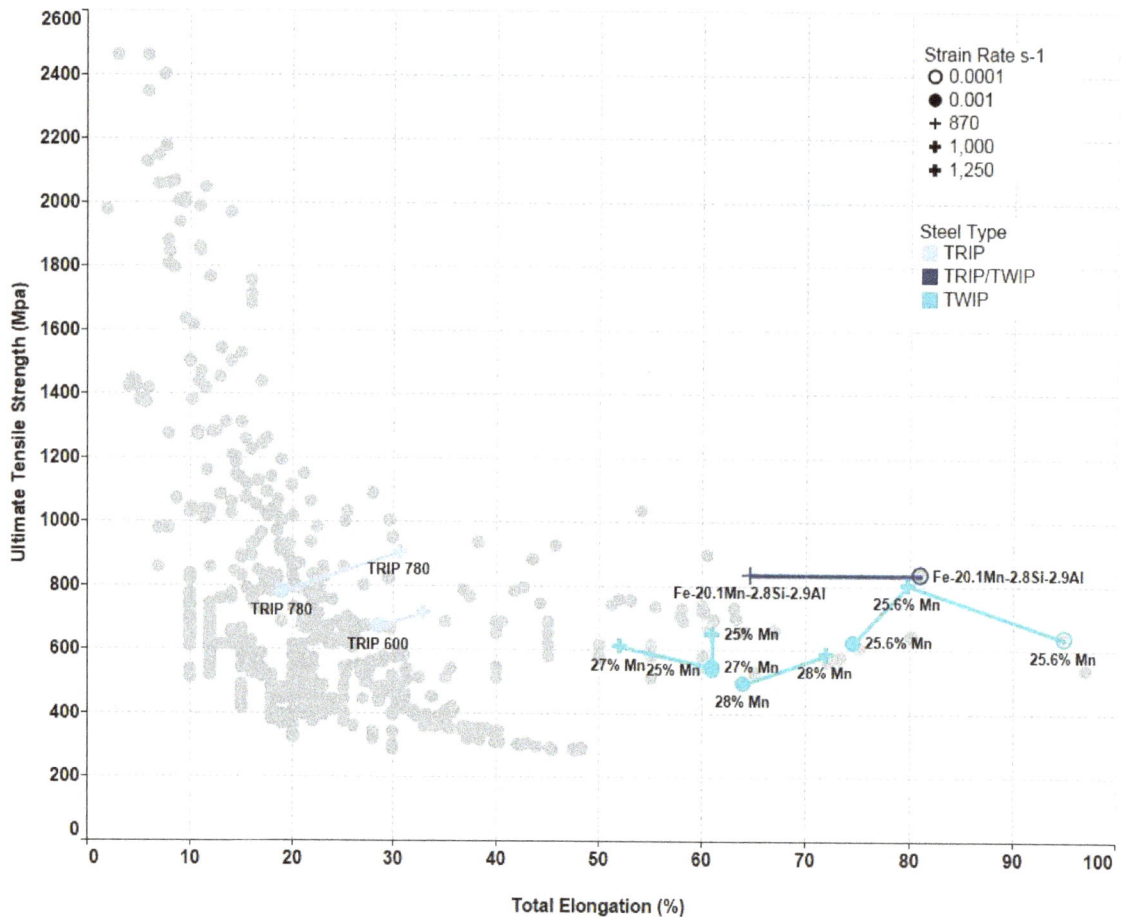

The data points used in 3 were obtained from reports by Curtze and Kuokkala (2010), Frommeyer et al. (2003), World Auto Steel website (Institute 2014), and Grässel et al. (2000).

Data obtained from the World Auto Steel website for TRIP 600 and TRIP 780 steel indicate that both the ultimate tensile strength and ductility increase at higher strain rates. In this study, TRIP 600 was less strain-rate sensitive that TRIP 780.

Khosravifard et al. (2013) studied the dynamic torsional properties of a 20.1% Mn TRIP/TWIP steel using a modified torsional SHB. When subjected to a strain rate of 516 s^{-1}, the reported true strain at failure was almost half of what had been reported under quasi-static conditions (3.2×10^{-3} s^{-1}). However, when subjected to a strain rate of 1709 s^{-1}, the maximum true stress was approximately 8% less and the true strain was approximately 16% less than that of the specimen loaded under quasi-static conditions. The results of this testing are shown in Figure 24.

Frommeyer et al. (2003) also investigated the strain-rate-dependent behavior of a purely austenitic, 25.6% Mn TWIP. They reported that total elongation to failure decreased sharply from strain rates of 10^{-4} s^{-1} to 10^{-1} s^{-1}, while the ultimate tensile strength was relatively unchanged at these strain rates. Higher strain rate testing was conducted using flywheel and TSHB techniques. The elongation to failure at a strain rate of 10^{-3} s^{-1} was comparable to the elongation reported for a strain rate of 10^3 s^{-1}, and the ultimate tensile strength increased by more than 25% over the same range of strain rates.

Figure 24. True stress-strain curves from Khosravifard et al. (2013).

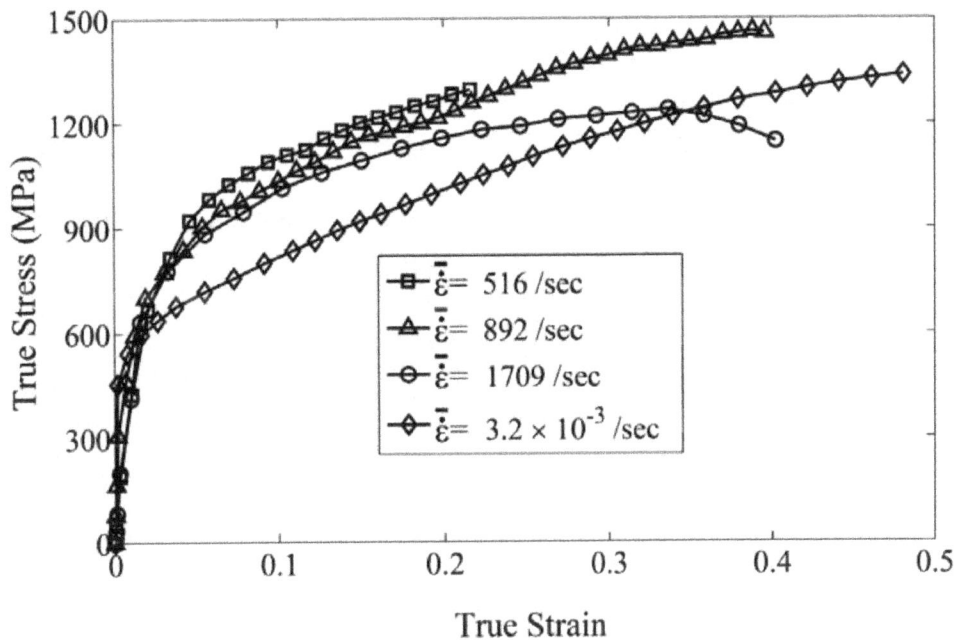

Grässel et al. (2000) studied an array of TRIP and TWIP steels with Mn contents ranging from 15.9% to 30.6% at strain rates between 10^{-4} s^{-1} and 10^3 s^{-1}. High strain rate testing was conducted using a flywheel apparatus. Under these conditions a 26.5% Mn TWIP steel displayed more strain-rate sensitivity than a 20.1% Mn TRIP/TWIP steel subjected to the same loading conditions. The ultimate tensile strengths of these two materials were almost the same at a strain rate of 10^3 s^{-1} despite the 26.5% Mn TWIP steel having an ultimate tensile strength that was 25% less than the 20.1% Mn TRIP/TWIP under quasi-static (10^{-4} s^{-1}) loading conditions. In general, the 20.1% Mn TRIP/TWIP steel had a very low strain rate sensitivity in both tension and compression.

Grässel et al. (2000) also noted that no martensite phase transformation was observed in the 26.5% Mn TWIP steel up to strain rates of 1.5×10^3 s^{-1}. Through x-ray defraction, only twinning behavior was observed.

Sahu et al. (2010) examined strain rate effects on the compressive mechanical properties of a 24.1% Mn, 0.91% Al TWIP metal and a 23.9% Mn, 3.5% Al TWIP metal. High strain-rate testing was conducted using a SHPB. Strain rates ranging from 10^{-4} s^{-1} to 4×10^3 s^{-1} were used, and Sahu et al. observed that the compressive true stress and true strain values reported at a strain rate of 1.2×10^3 s^{-1} was considerably smaller than quasi-static and higher strain rate values for both metals. These compressive true stress-strain curves are in Figure 25 (Sahu et al. 2010). This sort of behavior indicates that the mechanical properties of these two TWIP steels are not only highly strain rate dependent, but that the relationship is neither constant nor linear for strain rates between 10^{-4} s^{-1} and 1.2×10^3 s^{-1}.

Figure 25. Compressive true stress-strain curves from Sahu et al. (2010).
a) 24Mn-1Al TWIP steel
b) 24Mn-3Al TWIP steel

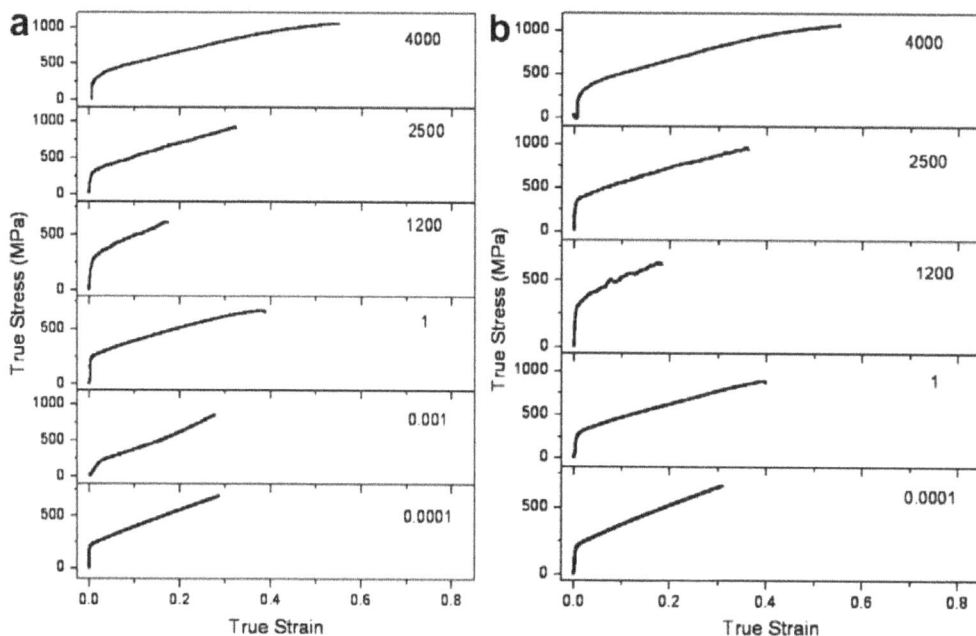

Curtze and Kuokkala (2010) observed that three TWIP steels with manganese contents ranging from 25-28% by weight exhibited only small increases in ultimate tensile strength from the quasi-static strain rate of 10^{-3} s^{-1} to the intermediate strain rate of 7.5×10^2 s^{-1}. However, these materials were much more strain-rate sensitive when tested at a strain rate of 10^3 s^{-1}, and ultimate tensile strengths at this high strain rate tended to be 10-20% higher than the strengths measured under quasi-static conditions. This rapid increase in ultimate tensile strength was attributed to viscous drag effects on the moving dislocations.

Curtze and Kuokkala (2010) also observed that while the ductility of one of the TWIP steels increased, another remained unchanged, and yet another decreased when subjected to high strain-rate loading conditions (1250 s^{-1}) as compared to quasi-static conditions (10^{-3} s^{-1}). This behavior was attributed to the stacking fault energy (SFE) of the materials, with the material possessing the highest stacking fault energy decreasing in ductility under high strain-rate conditions. The authors believed that the high SFE prevented the formation of mechanical twins critical to TWIP behavior.

Huh et al. (2008) reported that the elongation to failure remained constant for TRIP800 steel and the elongation to failure for TRIP600 steel increased by approximately 20% when subject to tensile loads applied at strain rates ranging from 0.003 to 200 s^{-1}.

Kim et al. (2013) developed model parameters for stress-strain behavior of TRIP 780 steels at strain-rates from .001 to 200 s^{-1}.

2.4 Stainless steels

2.4.1 Overview

Stainless steels can possess mechanical properties comparable to TRIP and TWIP steel. However, unlike the manganese-alloyed TRIP and TWIP steels, stainless steels are often alloyed with 12 to 35 wt. % chromium as well as smaller additions of nickel and silicon. These alloy additions produce steel with excellent corrosion resistance that can be attributed to a passive oxide film that forms in the presence of oxygen.

The ductility of these stainless steels is particularly high when the steels are mostly austenitic. For applications where large deformations can be sustained, these steels show great potential for energy absorption. Additionally, many martensitic stainless steels are high in tensile strength while still having some ductility and corrosion resistance. Figure 26 shows the quasi-static strength and elongation data for stainless steels.

Figure 26. Quasi-static ultimate tensile strength and total elongation properties for stainless steels.

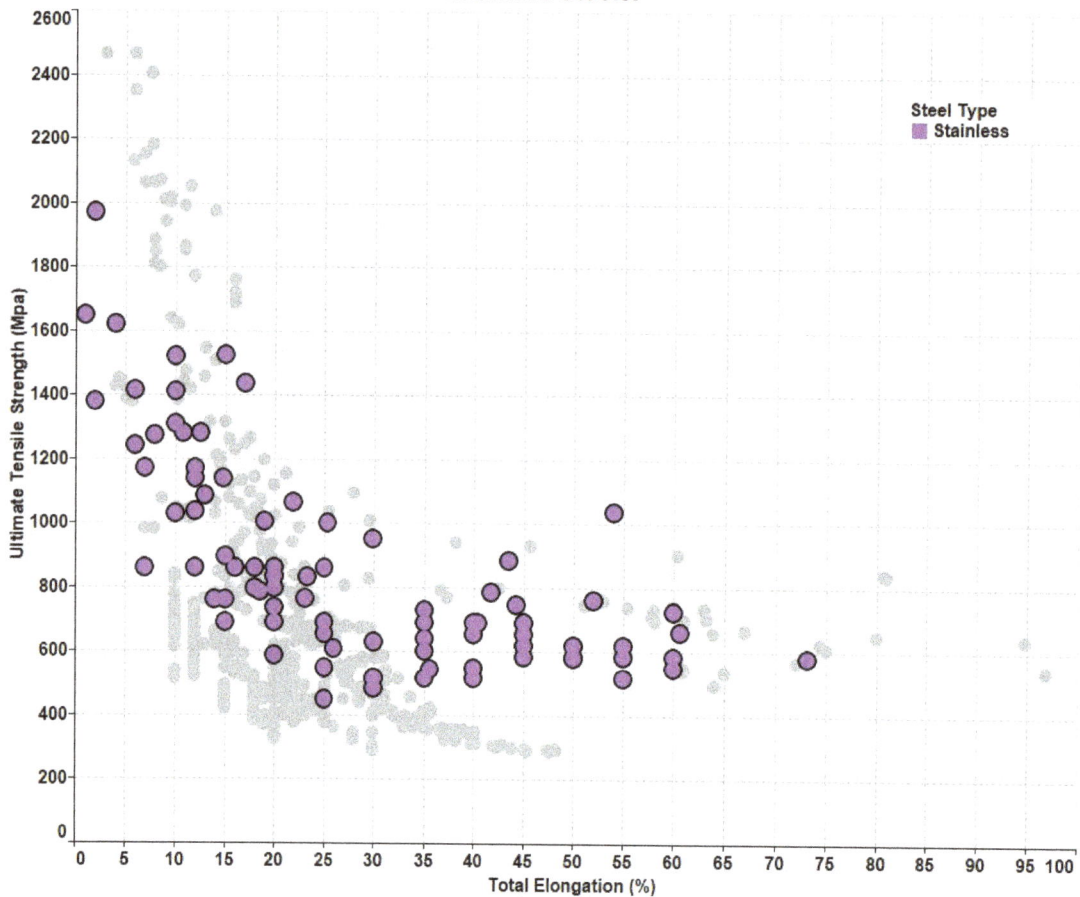

The data shown in Figure 26 were obtained from Davis (1998), Harvey (1992), the World Auto Steel website (Institute 2014), Milad et al. (2008), Byun et al. (2004), and De et al. (2006).

2.4.2 Classes of stainless steel

2.4.2.1 Martensitic

Martensitic stainless steels typically contain between .07 and 1.2 wt. % carbon and 12-18 wt. % chromium. The carbon content allows the martensite to transform into the austenite phase when heated. The martensite phase is very hard and strong but low in ductility. These steels have a distorted bcc crystalline microstructure and are the least corrosion-resistant grade of stainless steel. This ferromagnetic phase is very hard and strong but low in ductility. It can be hardened by heat treating and rapidly cooling (quenching). The most commonly used grade is AISI 410, although AISI 420 is commonly used to produce knives (Cardarelli 2008; Davis 1998).

2.4.2.2 Ferritic

Ferritic stainless steels typically have been alloyed to less than 0.2 wt. % carbon and 17-30 wt. % chromium. This lower carbon content prevents the formation of substantial amounts of the martensite phase. For a set ductility, ferritic stainless steels have slightly lower ultimate tensile strengths than other stainless steels, although they have an ultimate tensile strength approximately 50% greater than comparable plain carbon steels. Ferritic stainless steels are most commonly used for applications where a steel with moderate corrosion resistance is important, especially under conditions where stress-induced corrosion is likely to occur. Ferritic stainless steels are ferromagnetic, have a regular bcc crystalline microstructure, and cannot be hardened by heat treatment. However, they can be hardened by cold-working. The most common grade is AISI 430 (Cardarelli 2008; Davis 1998).

2.4.2.3 Austenitic

Austenitic stainless steels are the most commonly produced stainless steels. The face-centered cubic, iron-chromium microstructure of austenitic stainless steel is stabilized by nickel or manganese, and the carbon content is typically less than 0.15 wt. %. It is common for austenitic stainless steels to be alloyed with 8-20 wt. % nickel and 18-25 wt. % chromium. Austenitic steels are not hardened by heat treatment but can be hardened through cold-working. These steels are sought for their superior corrosion resistance. The most common austenitic stainless steel grade is AISI 304. Some grades such as 316 have molybdenum added for additional corrosion resistance. These steels should not be used in applications where temperatures may exceed 450°C (Cardarelli 2008; Davis 1998; Dini et al. 2010; Newson and Abp 2001; Hamada et al. 2013; Bouaziz and Barbier 2013).

2.4.2.4 Duplex

High chromium content, low nickel content stainless steels are sometimes classified as duplex. These steels have a microstructure that is a combination of ferrite and austenite. However, the tensile strength of duplex stainless steel is greater than either ferritic or austenitic stainless steels while still exhibiting great corrosion resistance. It is common for duplex stainless steels to be alloyed with 18-26 wt. % chromium, 4-7 wt. % nickel, and 2-3 wt. % molybdenum (Cardarelli 2008; Davis 1998).

2.4.2.5 Precipitation Hardened (PH)

PH steels occur when martensite undergoes a low-temperature heat treatment (aging). This treatment precipitates hard compounds and tempers the martensite, but this process is slowed in low carbon (L) steels. PH steels are very high strength, are not as corrosion-resistant as other stainless steels, yet may still be alloyed with 12-30 wt. % chromium (Cardarelli 2008; Davis 1998).

2.4.3   AISI/SAE designations

200-299   Chromium-Nickel-Manganese (austenitic).

300-399   Chromium-Nickel (austenitic).

400-499   Chromium (Martensitic and ferritic).

500-599   Low Chromium.

2.4.4   Material properties

Material properties for traditional stainless steel are summarized in Table 5.

Table 5. Material properties of traditional stainless steel (Cardarelli 2008).

Property	Value
Specific Gravity	7.6-7.8
Elastic Modulus (GPa)	200-215
Shear Modulus (GPa)	80-83
Poisson's Ratio	0.27-0.29
Coefficient of linear Thermal Expansion α: $\left(\frac{1}{K} \cdot 10^{-6}\right)$	9.3-10.8
Thermal Conductivity κ: $\left(\frac{W}{K \cdot m}\right)$	20.9-27.0
Electrical Resistivity ρ: $(\mu\Omega \cdot cm)$	60-67

*Stainless steels are sometimes alloyed with nitrogen or boron to increase strength and corrosion resistance.

Many stainless steels have mechanical properties that are greatly reduced under high thermal conditions and, therefore, are not suited for applications that may be exposed to elevated temperatures. However, Cast H-series stainless steels are higher in nickel and carbon than other stainless steels and can be used in conditions where thermal exposure is between 650 and 1200°C.

2.4.5   Quasi-static tensile properties

As is shown in Figure 27, martensitic stainless steels generally have a higher ultimate tensile strength and lower ductility than ferritic and austenitic steels. As is shown by Figure 28 and Figure 27, austenitic stainless steels have the greatest ductility of the stainless steels.

Milad et al. (2008) and De et al. (2006) examined the impact that cold-working has on the mechanical properties of a 304 austenitic stainless steel (Figure 29). During the cold-working process, (γ) austenite is transformed to (α') martensite. The stress induced formation of these phases gives rise to the notion of transformation-induced plasticity (TRIP). Byun et al. (2004) also found similar stress-induced behavior in a 316LN austenitic stainless steel.

Byun et al. (2004) also found that the mechanical properties of 300 series austenitic stainless steels are highly dependent on the temperature at the time of testing. In these experiments, ductility decreased by as much as 45% when the testing temperature was increased from -50°C to 100°C.

The data shown in Figure 28 and Figure 27 were obtained from Davis (1998), Harvey (1982), the World Auto Steel website (Institute 2014), Milad et al. (2008), Byun et al. (2004), and De et al. (2006).

Figure 27. Quasi-static ultimate tensile strength and total elongation properties for martensitic and precipitate hardened stainless steels.

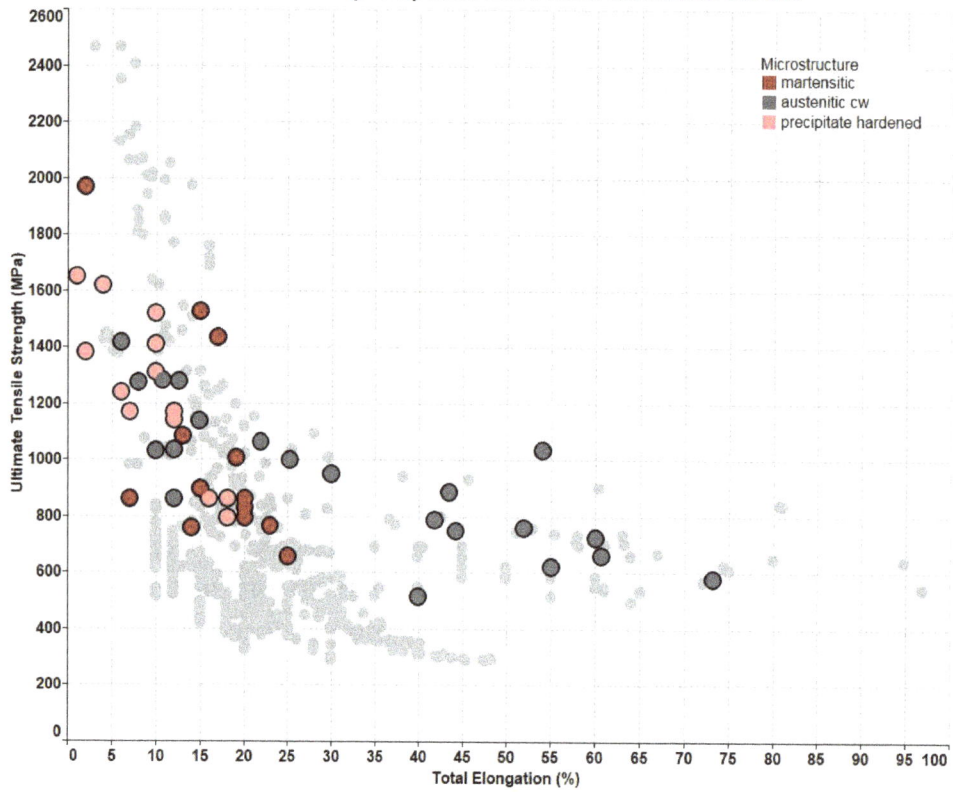

Figure 28. Quasi-static ultimate tensile strength and total elongation properties for austenitic, ferritic, and duplex stainless steels.

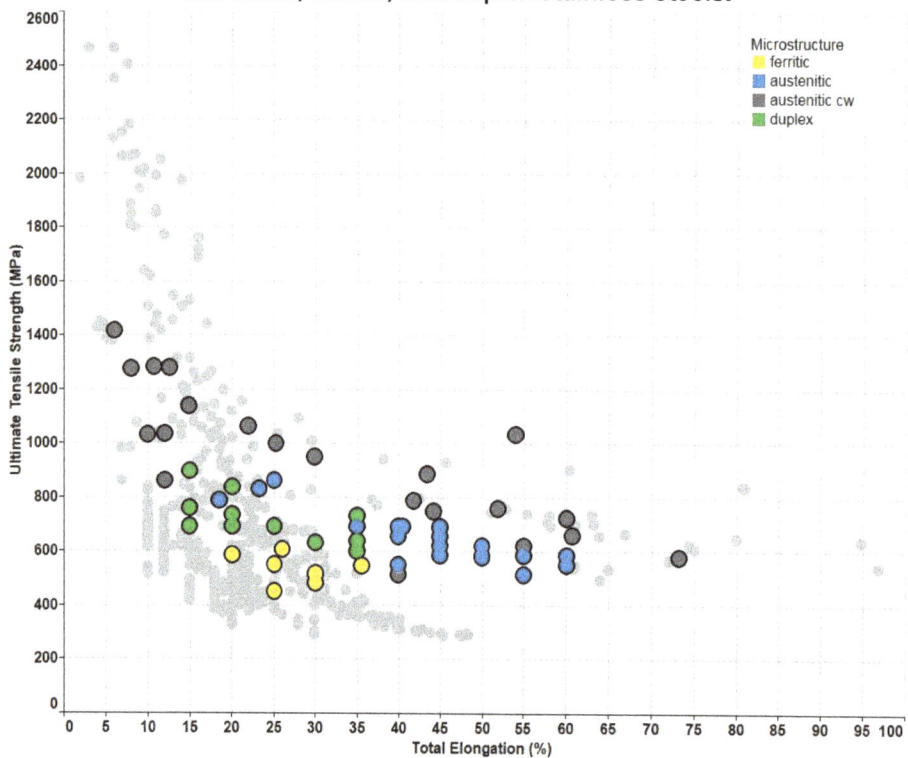

The data shown in Figure 29 were obtained from Davis (1998), Harvey (1982), and Milad et al. (2008).

Figure 29. Effect of cold-working austenitic stainless steel.

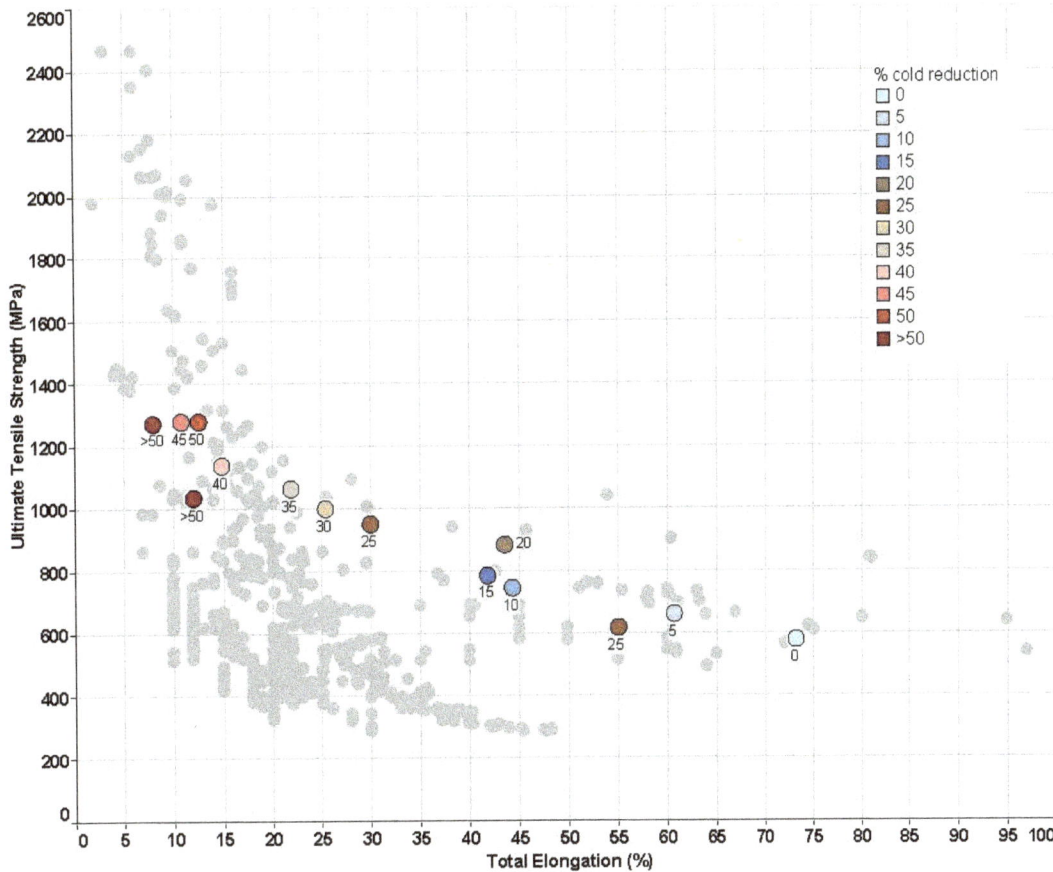

Das et al. (2008) found that the ductility of an austenitic 304LN stainless steel was highly strain rate dependent at strain rates of .0001 to 1.0 s^{-1}. This variation is shown in Figure 30.

Figure 30. Quasi-static and intermediate strain-rate dependence of a 304 LN austenitic stainless steel from Das et al. (2008).

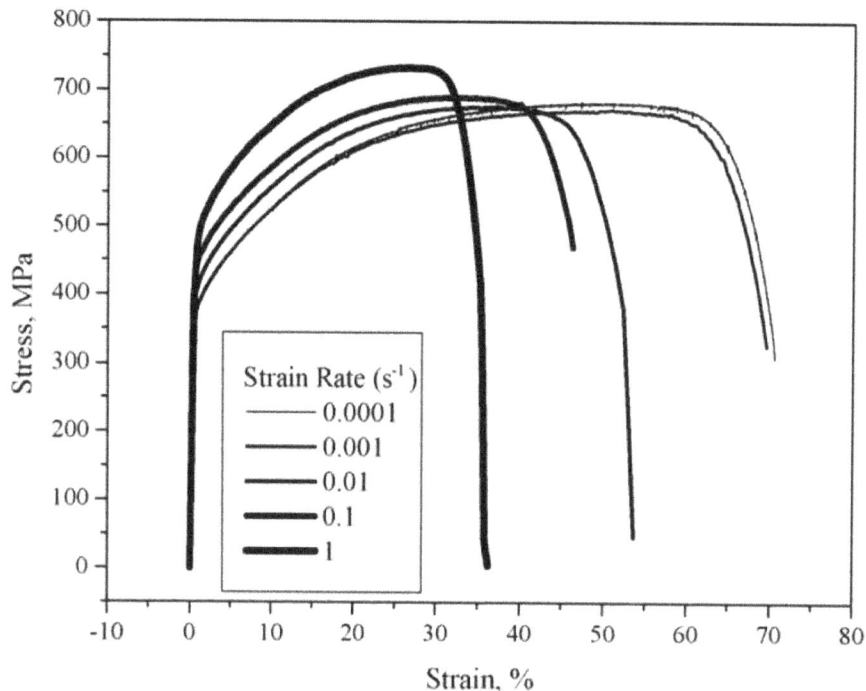

Guan et al. (2013) studied the use of selective laser melting (SLM) technology for manufacturing a 304 austenitic stainless steel and reported that the ultimate tensile strength, yield strength, and elongation to failure is dependent on the orientation of the laser and powder bed.

2.4.6 Dynamic tensile properties

Dynamic testing of austenitic grades 304L and 321 showed that the ultimate tensile strength of the materials exhibited moderate to high strain rate dependence, while the ductility of the material was relatively unaffected by strain rate (Figure 31). The data shown in Figure 31 were retrieved from the World Auto Steel website (Institute 2014).

Lee et al. (2010) tested 304L austenitic stainless steel specimens at very high strain rates using a compressive SHPB and found that, when strain rates exceeded 4000 s^{-1}, little austenite was transformed to martensite. Also, Lee et al. (2010) postulated that the relationship between martensite formation and strain rate may be linear for strain rates between 2000 and 6000 s^{-1} under compressive testing conditions.

Figure 31. Dynamic and quasi-static mechanical properties of stainless steel.

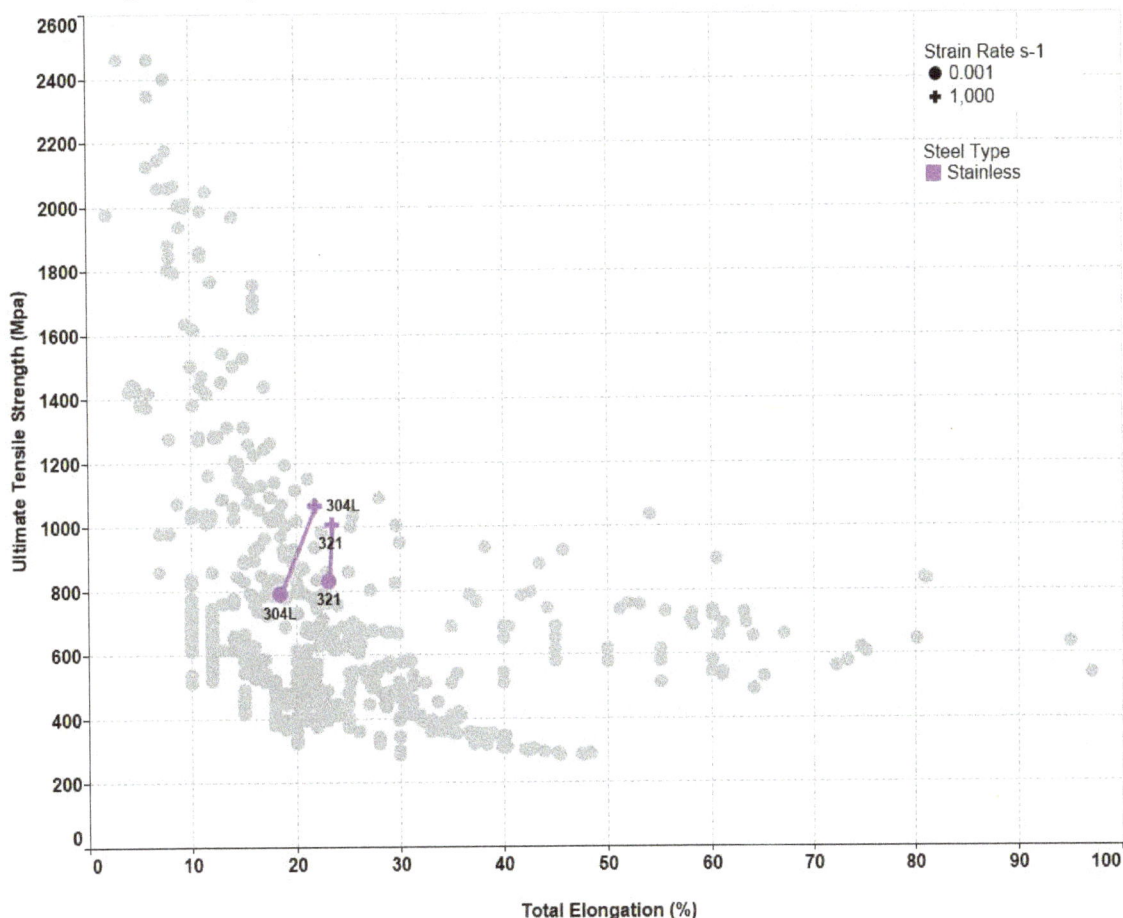

Guo and Nemat-Nasser (2006) developed a "physically based" constitutive model that correlated to the compressive stress-strain behavior of Nitronic®50, a highly corrosion-resistant, nitrogen- and boron-hardened, austenitic stainless steel. The Nitronic®50 was tested under a variety of thermal conditions ranging from 77 K to 1000 K and a variety of strain rates ranging from 0.001 s^{-1} to 8000 s^{-1}. The model correlates closely to these experimental data but does not account for dynamic strain aging effects, which, according to the author, are important for tests conducted between strains rates of 0.001 s^{-1} and 0.1 s^{-1} and at temperatures of 400-1000 K.

2.5 High-performance alloy steels

2.5.1 Introduction

High-performance (HP) alloy steels have ultimate tensile strengths that exceed 1000 MPa while still exhibiting some ductility. Most of these steels were developed for specific purposes such as submarine armor, turbine blades, or other applications where high strength or high energy absorption capacity is necessary and large deformations cannot be sustained (Davis 1998; Bardelcik et al. 2010, 2012). Figure 32 shows the quasi-static strength and elongation data for high-performance alloy steels.

Many of the materials have complex processing regimens and chemical compositions that consist of many alloys. These features are specific to each material. However, the high alloy contents of many of these steels have made it possible to have a very high strength steel with moderate corrosion resistance.

Figure 32. Quasi-static ultimate tensile strength and total elongation properties for high-performance alloy steels.

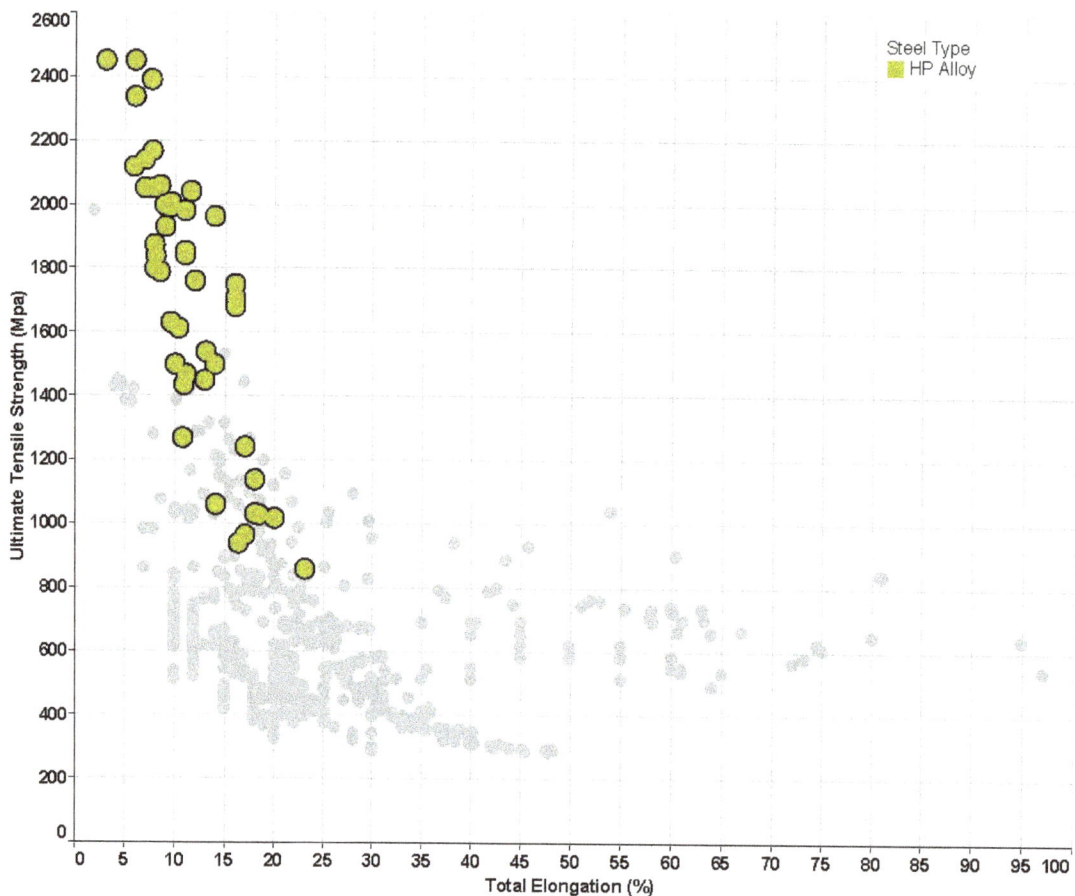

The data shown in Figure 32 were obtained from Davis (1998), Harvey (1982), and Bardelcik et al. (2010, 2012).

Steels that fall into this category are usually proprietary materials, each with its own unique heat treatment, processing method, and chemical composition. Many materials utilize technologies such as vacuum degassing and refrigeration of steel at temperatures of -73°C or cooler during the production process (Davis 1998).

2.5.2 Quasi-static tensile properties

The quasi-static mechanical properties of steels in this category are highly dependent on the heat treatments and processing methods used to produce the steel. As is shown in Figure 33, the way in which the steel is produced has a greater effect on quasi-static mechanical performance than on how the material is classified.

Maraging steels are typically high in nickel, cobalt, and molybdenum and very low in carbon. Their high strength is attributed to an aging process during which hardened intermetallic phases precipitate at temperatures near 480°C (Davis 1998).

AISI/SAE 4340 steel has good fatigue resistance and is used to produce thin plates, aircraft landing gear, and bolts. It has problems with stress-induced corrosion cracking especially in marine environments, which can be mitigated though nitriding (Davis 1998).

In comparison, 300M is similar to AISI/SAE 4340 steel except for the addition of more silicon, carbon, molybdenum, and vanadium. The addition of these alloys results in a stronger, more difficult-to-produce steel. Like the 4340 steel, this material has also been used in aircraft landing gear and thin plates (Davis 1998).

Figure 33. Quasi-static ultimate tensile strength and total elongation properties for common high-performance alloy steel designations.

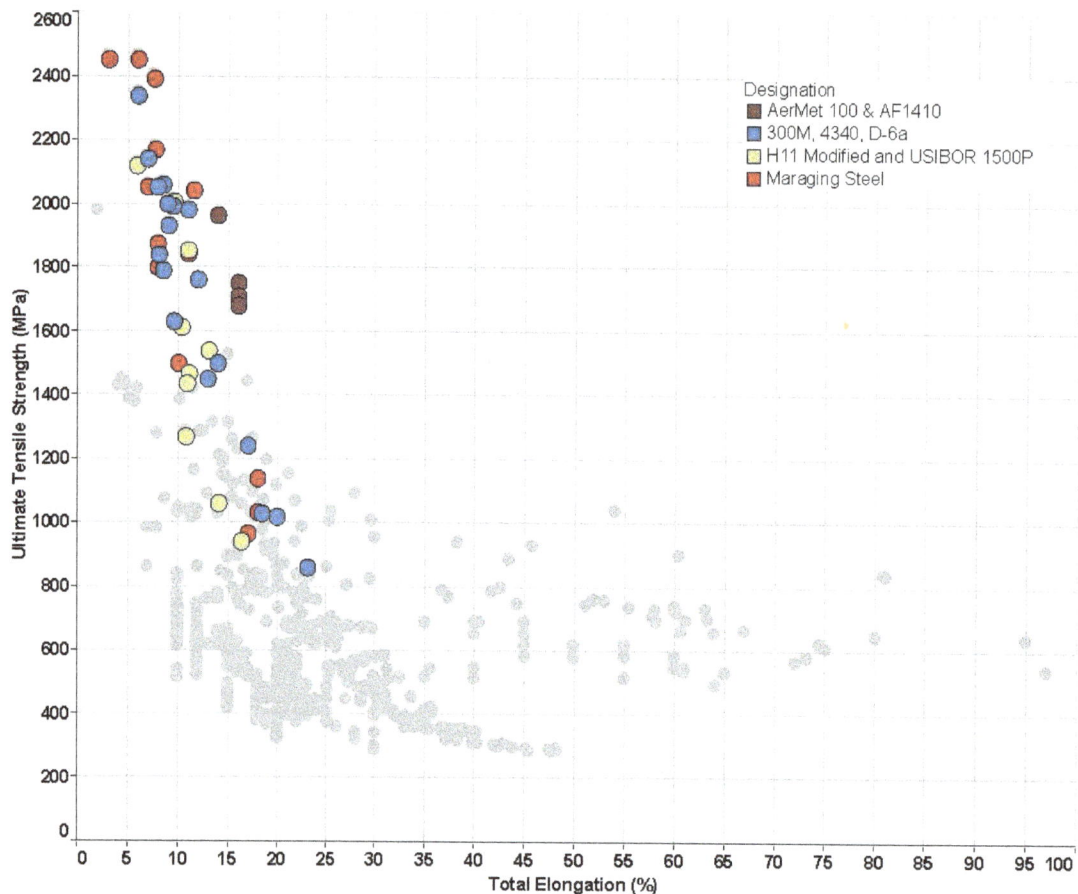

The data shown in Figure 33 were obtained from Davis (1998), Harvey (1982), and Bardelcik et al. (2010, 2012).

D-6a was developed for aerospace applications including structural aircraft components, motor cases for solid-fuel rockets, backer blocks, and thin plates. One production method involves melting material in an electric arc furnace then remelting the material with a vacuum arc. D-6a is equally susceptible to corrosion cracking as a comparable strength AISI/SAE 4340 steel (Davis 1998).

H11 modified steels are similar to traditional H11 tool steels, but different processing methods result in a more stable material under elevated thermal loading conditions. This material is considered easy to weld and is typically used in thin plates, aerospace applications, steam and gas turbines, and hot-work dies (Davis 1998).

H13 is similar to H11 steel in composition and mechanical performance. However, the addition of vanadium produces a more wear-resistant material (Davis 1998).

AF1410 was developed by the U.S. Air Force and has been used to produce submarine hulls and various aircraft structural components. It is an iron-nickel-cobalt alloy that exhibits corrosion resistance to marine environments. Continuous wave-tungsten arc welding is commonly used on this type of steel. The processing methods used to produce this material are complex (Davis 1998).

AerMet 100 was originally developed for use in armoring systems and jet engine shafts. It is a high-strength, nickel-cobalt steel that was patented in 1991 (Davis 1998).

USIBOR 1500P was designed for use in automobile structural and safety components. It has been alloyed with manganese and boron, and the surface is often covered by a galvanized coating to improve corrosion resistance (Bardelcik et al. 2010, 2012).

2.5.3 Dynamic tensile properties

Bardelcik et al. (2010, 2012) investigated the tensile mechanical properties of a USIBOR 1500P metal that had been annealed at 900°C and quenched in various media. These changes in quenching media resulting in cooling rates between 25°C/s and 2200°C/s. The mechanical testing was conducted under quasi-static (.003 s^{-1}) and dynamic conditions (960 s^{-1}) using TSHB equipment. The USIBOR 1500P test specimens were cooled at different rates to examine the impact of various quenching media. A summary of reported ultimate tensile strength and elongation to failure data is shown in Figure 34.

Through the testing performed by Bardelcik et al., it was clear that the steel that had cooled at a rate of 25°C/s had a lower ultimate tensile strength and was more strain-rate sensitive than the steels that had cooled more quickly. The steel subjected to a cooling rate of 2200°C/s clearly had the greatest ultimate tensile strength and lowest strain rate sensitivity. However, this material experienced a slight loss in ductility when subjected to dynamic loading conditions, while metals subjected to other quenching media in this study experienced an increase in ductility. The improvements in tensile strength achieved by quenching the steel at different rates appear to be more pronounced under quasi-static loading conditions than at high strain rates when tested in tension.

Figure 34. Dynamic and quasi-static mechanical properties of high-performance alloy steel.

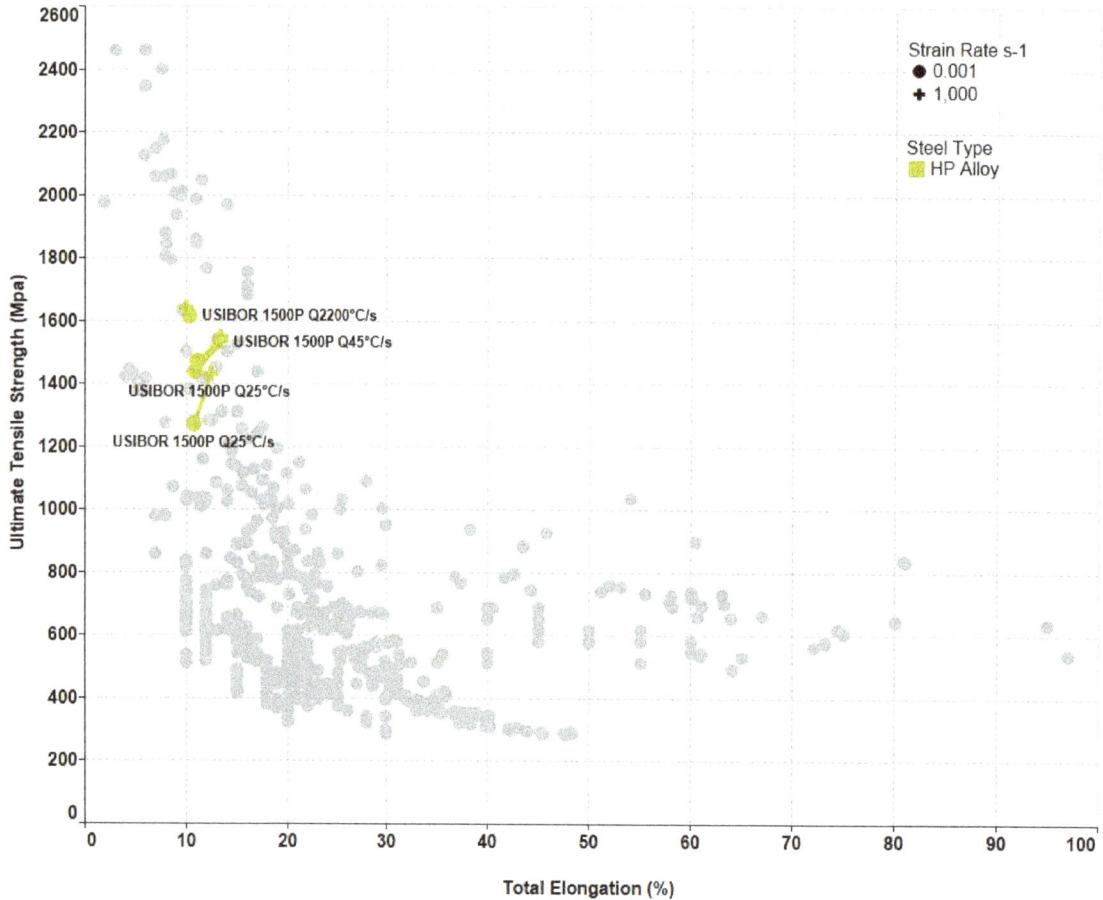

The data points shown in Figure 34 were obtained from reports written by Bardelcik et al. (2010, 2012).

2.5.4   Existing model data

Bardelcik et al. (2012) found that the Johnson-Cook and Zerilli-Armstrong constitutive models were unable to account for the hardening behavior resulting from samples being quenched in various media. Therefore, a modified version of the Voce hardening model was used to model the hardening produced by the quenching process and the strain-rate dependent behavior.

3   Aluminum Alloys

3.1   Overview

Aluminum metals are considerably lower in strength than iron, nickel, cobalt, and titanium. However, depending on the application, aluminum metals may be a lightweight, economical alternative to other metals. Consequently, aluminum alloys have been extensively researched by the automotive and aerospace industries.

Aluminum metals on average have a lower strain-rate sensitivity than other metals. However, several authors reported that the tensile strength of aluminum decreased at high strain-rates when compared to quasi-static testing (Hadianfard et al. 2008; Smerd et al. 2005; Mukai et al. 1995).

3.2   Common AISI/SAE designations

- 1xxx-wrought aluminum containing greater than 99.00 wt. % aluminum.
- 1xx.x-cast aluminum containing greater than 99.00 wt. % aluminum.
- 2xxx-wrought aluminum metals containing copper and manganese alloys. Used to produce aircraft parts, truck wheels, and hinges; this class of aluminum alloys is relatively high strength and has great machinability and workability (Cardarelli 2008).
- 2xx.x-cast aluminum metals containing copper alloys.
- 3xxx-wrought aluminum metals containing manganese alloys. Alloy 3003 is commonly used to manufacture ductwork and garage doors. Higher strength than 1100 Al (Cardarelli 2008).
- 3xx.x-cast aluminum metals containing silicon, copper, and manganese alloys.
- 4xxx-wrought aluminum metals contain silicon alloys.
- 4xx.x-cast aluminum metals containing silicon alloys.
- 5xxx-wrought aluminum metals containing magnesium alloys. These alloys are finer grained than 3xxx aluminum but have comparable mechanical properties. Therefore, these alloys are typically used to reduce finishing cost for applications where surface roughness is a concern (Cardarelli 2008).
- 5xx.x-cast aluminum metals containing magnesium alloys.
- 6xxx-wrought aluminum metals containing magnesium and silicon alloys. Alloy 6061 is commonly used to manufacture the bodies of trucks, boats, and busses. It is also used for general structural applications (Cardarelli 2008).

- 6xx.x-cast aluminum metals containing magnesium and silicon alloys.
- 7xxx-wrought aluminum metals containing zinc and magnesium alloys sometimes also containing copper as an alloy. Alloy 7075 is commonly used for aircraft parts where exceptionally high strengths are necessary (Cardarelli 2008).
- 7xx.x-cast aluminum metals containing zinc.
- 8xxx-wrought aluminum containing alloying elements not covered in other designations.
- 8xx.x-cast aluminum metals containing tin alloys.
- 9xx.x-cast aluminum metals containing alloying elements not covered in other designations.

Temper designations:

F-as manufactured

O-annealed

H-strain hardened (wrought alloys)

W-solution heat treated

T-thermally treated (other)

3.3 Material properties

Material properties for aluminum metals and alloys are summarized in Table 6.

Table 6. Material properties for aluminum metals and alloys (from Cardarelli 2008, Mukai et al. 1995).

Property	Value
Specific Gravity	2.5-2.83 (pure Al 2.71)
Coefficient of linear Thermal Expansion $\alpha: \left(\frac{1}{K} \cdot 10^{-6}\right)$	18-24 (pure Al 23.6)
Thermal Conductivity $\kappa: \left(\frac{W}{K \cdot m}\right)$	93-234 (pure Al 222-234)
Electrical Resistivity $\rho: (\mu\Omega \cdot cm)$	2.7-6.9 (pure Al 2.7-3.0)

Note that aluminum metals tend to dissolve when exposed to corrosive solutions such as hydrochloric acid, sulfuric acid, and sodium hydroxide. A common byproduct of this chemical reaction is hydrogen gas. Likewise, when aluminum is exposed to fresh concrete in the field, top layers of aluminum will react with the highly alkaline concrete. The aluminum will eventually develop a passive layer that will provide some protection against the alkaline solution. However, the hydrogen gas that is released by this chemical reaction can negatively impact concrete properties and bond strength and cause expansion of the fresh concrete that may endanger formwork. For these reasons, care must be taken in creating any composite system where aluminum and concrete will come into contact.

3.4 Quasi-static tensile properties

Aluminum metals are lower in tensile strength than other metals included in this report and do not show an improvement in ductility. Figure 35 shows how the ultimate tensile strength and elongation to failure of aluminum alloys compare to other metals included in this report. The primary benefits to aluminum metals are economy, machinability, and weight.

Figure 35. Quasi-static ultimate tensile strength and total elongation properties for aluminum-based alloys.

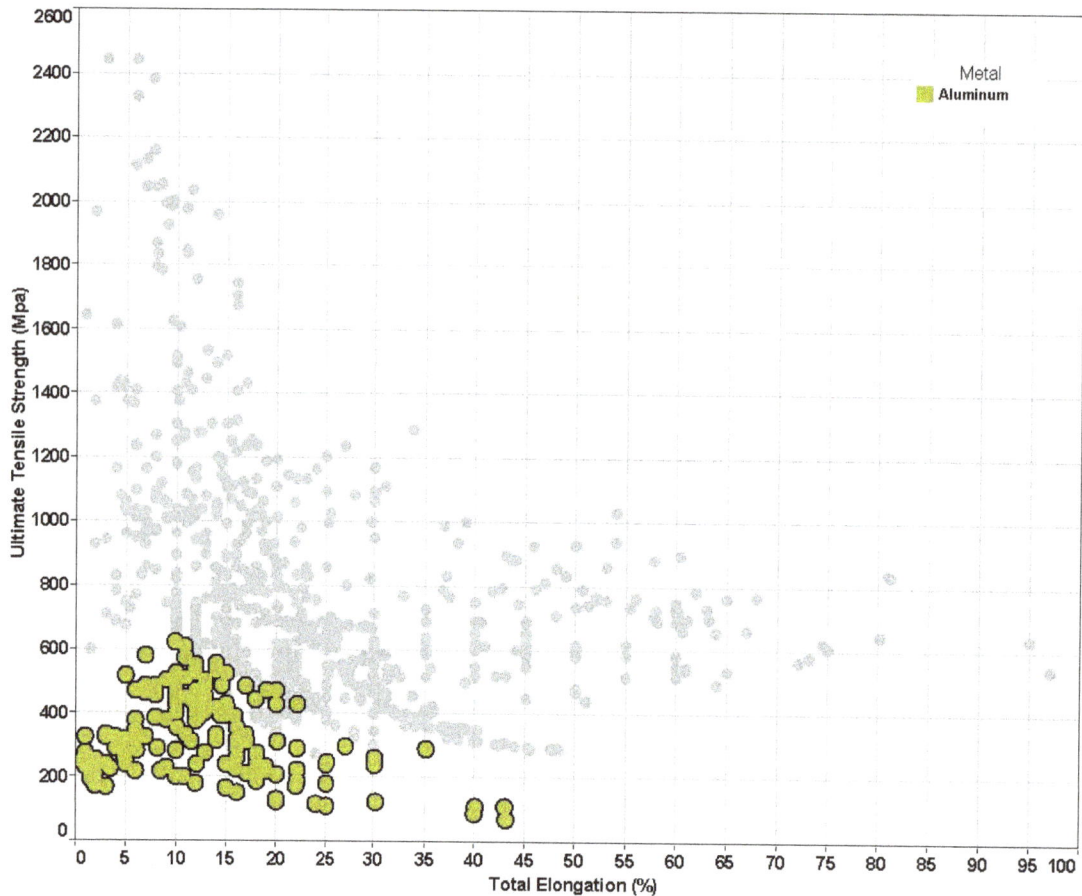

The data shown in Figure 35 were obtained from the ASM online alloy center database (2015).

3.5 Dynamic tensile properties

Smerd et al. (2005) conducted a series of tests to examine the mechanical properties of AA5182 and AA5754 sheet metal. These properties were measured using dog-bone shaped specimens that were tested at quasi-static and dynamic strain rates using an Instron testing machine and TSHB, respectively.

Both metals had a quasi-static, total elongation to failure of about 30%. Under TSHB test conditions at a strain rate of 1500 s⁻¹, AA5754 had a total elongation to failure of 43.5%, and AA5182 had a total elongation to failure of 37.5%. In general, AA5754 was more strain-rate sensitive in this study than AA5182. The Johnson-Cook constitutive model was also examined for both metals in this study (Smerd et al. 2005).

Hadianfard et al. (2008) collaborated with Smerd et al. (2005) to expand on the data set gathered for AA5754 and AA5182. The resulting article includes more test data about the mechanical properties of these two sheet metals at different quasi-static strain rates, and it adds SEM analysis of damage mechanisms at both quasi-static and dynamic strain rates. Few, if any, shear bands were observed on the TSHB specimens that were tested at room temperature. If shear banding was present in the TSHB specimens, this region did not exhibit microstructural damage including voids, microcracks, and damaged second phase particles. Any damage observed was very close to the fracture surface. The primary mode of failure in the dynamic test specimens appeared to be void nucleation, growth, and coalescence prior to the development of significant shear bands.

This was in sharp contrast to the multiple large shear bands observed on specimens tested under quasi-static conditions. The specimen microstructure included a large number of voids, microcracks, and damaged second phase particles, each of which were concentrated within the shear bands. Shear localization appeared to be the primary mode of failure in these test specimens (Hadianfard et al. 2008).

Kozmel et al. (2015) studied the high strain-rate compressive mechanical properties of Al-Cu, Al-Cu-Mn-Mg, and AA2139 alloys that were tested using a SHPB. They found that, as the alloy amount increased, the compressive strength also increased. However, the aluminum alloy with the greatest alloy content (AA 2139) was much less ductile. The increase in compressive strength was believed to be due to misorientation of the microstructure because of the presence of the alloying elements. This microstructure forced an irregular slip surface that provided resistance to shear banding.

Mukai et al. (1995) examined the effect of grain size on the mechanical properties of an IN905XL aluminum alloy. The grain sizes of the alloy were varied by extruding it at different temperatures. Five grain sizes ranging from 0.8 μm to 8.1 μm were studied under both quasi-static and dynamic strain rates using an Instron testing machine and a modified TSHB, respectively. The results of this testing are shown in Figure 36.

Figure 36. Typical stress-strain curves for an IN905XL aluminum alloy, produced with five different grain sizes (from Mukai et al. 1995). The mean grain sizes were approximately 0.8 μm (FG), 1.4 μm (IG1), 1.9 μm (IG2), 4.3 μm (CG1), and 8.1 μm (CG2).

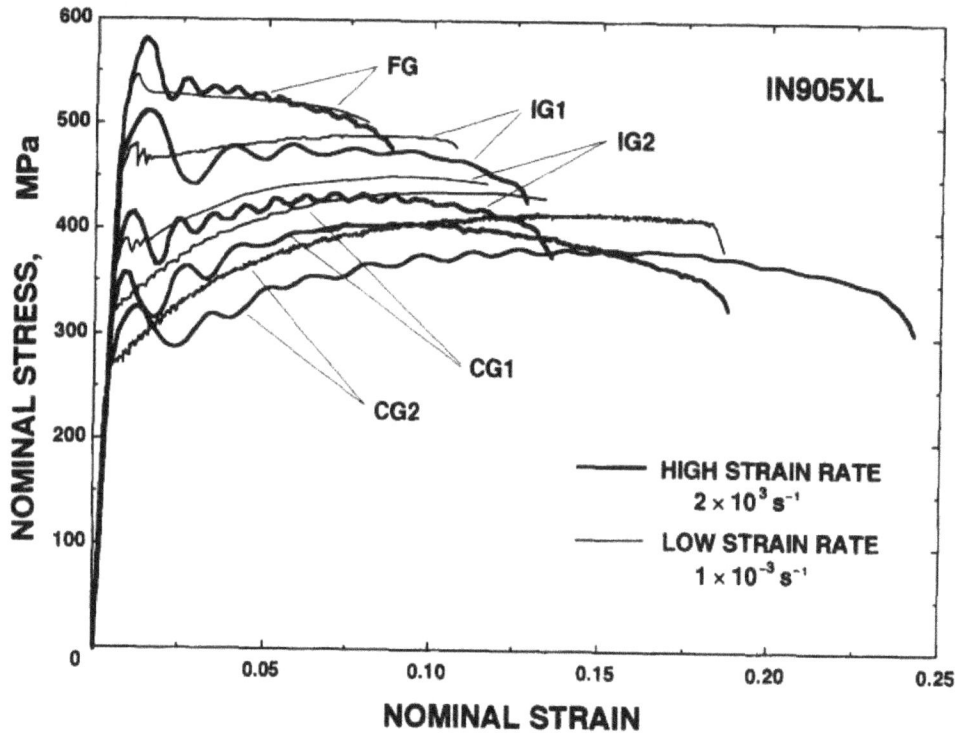

In general, finer-grained materials tend to be less strain-rate sensitive, higher strength, and less ductile than coarser-grained aluminum alloys (Hadianfard et al. 2008; Mukai et al. 1995). Many authors indicate that, although strain rate sensitivity is relatively low for many aluminum metals between strain rates of 10^{-3} s^{-1} and 10^3 s^{-1}, strain rate sensitivity increases dramatically at strain rates greater than 10^3 s^{-1} (Hadianfard et al. 2008; Smerd et al. 2005; Mukai et al. 1995; and Guo et al. 2011).

Chen et al. (2009) examined the dynamic tensile properties of AA6060-T6, AA6082-T6, AA7003-T6, and AA7108-T6 at strain rates ranging from 10^{-3} s^{-1} to 10^3 s^{-1}. Quasi-static and intermediate strain rate testing was conducted using a servo-hydraulic test machine, and high-rate testing utilized a TSHB. The orientation of the test specimen with respect to extrusion direction was also studied. In this study, AA6060-T6 and AA6082-T6 showed only slight positive strain rate sensitivity. However, the mechanical properties of AA7003-T6 and AA7108-T6 were much more strain-rate sensitive, with significant increases in both true stress and true strain at higher strain rates.

Tucker et al. (2010) studied the stress, strain, and microstructural deformation of rolled AA5083-H13, cast AA356-T6, and extruded AA6061-T6 alloys in tension, compression, and torsion at strain rates ranging from $10^{-3}\,s^{-1}$ to $1.5 \times 10^3\,s^{-1}$. All of the dynamic testing was conducted using various Split Hopkinson Bar test setups. The authors noted that the maximum flow stress for AA5083-H13 increased slightly at high strain rates, but the elongation to failure decreased slightly. Both the maximum flow stress and elongation to failure increased for AA6061-T6 at high strain rates for all 3 loading conditions. The AA356-T6 alloy exhibited very little strain rate sensitivity. Fractured AA356-T6 specimens were similar in appearance when tested under quasi-static and dynamic loading conditions. Conversely, notable necking was observed in both AA6061-T6 and AA5083-H13 during high strain-rate tensile testing, although the amount of necking was greater for the AA6061-T6 specimens.

3.6   Other dynamic testing

Melby and Eide (2013) conducted a series of tests on sheets of AA1050-H14 using a compressed gas gun (more similar to ballistic impact than blast impact loading). LS-DYNA and IMPETUS Afea models were then developed to describe the material behavior when subjected to this loading. However, the authors note that the peak pressure and impulse loadings generated using the gas gun are different from those achieved in a blast-type situation. During their study, they observed little difference in the amount of deformation between solid aluminum plates, aluminum plates containing small slits, and aluminum plates containing small, diamond-shaped notches.

Ghosh et al. (2015) studied the effect of natural and artificial aging on the dynamic compressive stress and strain of AA 2099. They found that artificial aging resulting from heating the samples at temperatures of 120-160°C for 12 to 18 hr caused an increase in the number of precipitates in the microstructure. These precipitates were initially irregularly distributed and became coarser and started to nucleate when aged for greater periods of time. The specimens were tested using a direct impact Hopkinson bar, where a blunt projectile was fired at the specimen having an impact momentum of 39 kg·m/s. The addition of these irregular, coarse precipitates in the artificially aged specimens caused these specimens to be considerably less ductile under dynamic loading conditions than the naturally aged specimens. The aging treatment had far less effect on the maximum flow stress observed in the material.

4 Nickel- and Cobalt-Based Alloys

4.1 Overview

Nickel- and cobalt-based alloys, or super alloys, are well-suited for corrosive environments subjected to large changes in temperature. The increased temperature and corrosion resistance, as well as the ability to withstand creep, makes nickel and cobalt super alloys ideal for a wide range of applications including cryogenic storage tanks, nuclear reactors, space equipment, and turbine blades (Amato et al. 2012, Byun and Farrell 2003, Chlebus et al. 2015, Kanagarajah et al. 2013, and Lee et al. 2011).

Nickel-based super alloys display a superior creep resistance at elevated temperatures, while Co-based super alloys display an increased hot corrosion, wear, and oxidation resistance. These properties have made Co-based super alloys of higher interests for recent research. Some common alloying elements are chromium (Cr), aluminum (Al), titanium (Ti), molybdenum (Mo), tungsten (W), niobium (Nb), tantalum (Ta), and cobalt (Co) (X. Zhang et al. 2014).

4.2 Material properties

Typical material properties for nickel alloy steels are in Table 7.

Table 7. Material properties of nickel alloy steels (Cardarelli 2008).

Property	Value
Specific Gravity	7.70-9.24 (pure Ni 8.89)
Elastic Modulus (GPa)	130-226
Coefficient of linear Thermal Expansion $\alpha: \left(\frac{1}{K} \cdot 10^{-6}\right)$	7.6-15.0 (pure Ni 13.3)
Thermal Conductivity $\kappa: \left(\frac{W}{K \cdot m}\right)$	8.9-24.0
Electrical Resistivity $\rho: (\mu\Omega \cdot cm)$	7.5-131 (pure Ni 7.5-9.5)

At room temperature, nickel is highly corrosion-resistant against alkaline solutions such as NaOH and KOH and generally does not corrode in moist environments. Nickel reacts readily with fluorine gas to form a protective passive layer of nickel fluoride. Thus, some alloys of nickel are commonly used to contain fluorine gas and hydrofluoric acid.

Inconel 718 is a common nickel-based super alloy and was developed for the high-temperature sections of aircraft jet engines and gas turbines for the aerospace and power industries. Inconel 718 displays a high wear and corrosion resistance as well as increased strength at elevated temperatures. Unfortunately, the alloy is difficult to machine due to the severe tool wear and poor final surface quality of the machined part (Thakur et al. 2009).

To promote high-temperature properties, different transition metals are used in the solid solution strengthening of Ni-based super alloys. Rare earth elements are instrumental in the increase of high-temperature creep resistance; however, they can also lead to the formation of topological close-packed phases decreasing the overall mechanical performance (Q. Zhang et al. 2014).

4.3 Quasi-static tensile properties

When tested under quasi-static strain rates, Ni-based super alloys display a significantly larger amount of elongation to failure over their Co-based counterparts. The data reveal that as the strength of Co-based super al-loys increases, the amount of ductility the materials exhibit significantly decreases. However, the nickel-based super alloys can obtain both increased strength and increased elongation to failure producing a material with high toughness.

Figure 37 shows that both the heat treatment and the chemical composition can greatly vary the mechanical properties of super alloys ranging from an elongation to failure of 1.5% to 66%. With such a large range of possible properties, Ni-super alloys can be engineered to fit most design needs where a high-strength ductile material needs to also withstand high temperatures. Co-super alloys are still in the early stages of development, and alloying has shown significant improvements to the overall ductility of the material without significant loss in strength.

Figure 37. Quasi-static ultimate tensile strength and total elongation properties for nickel- and cobalt-based alloys.

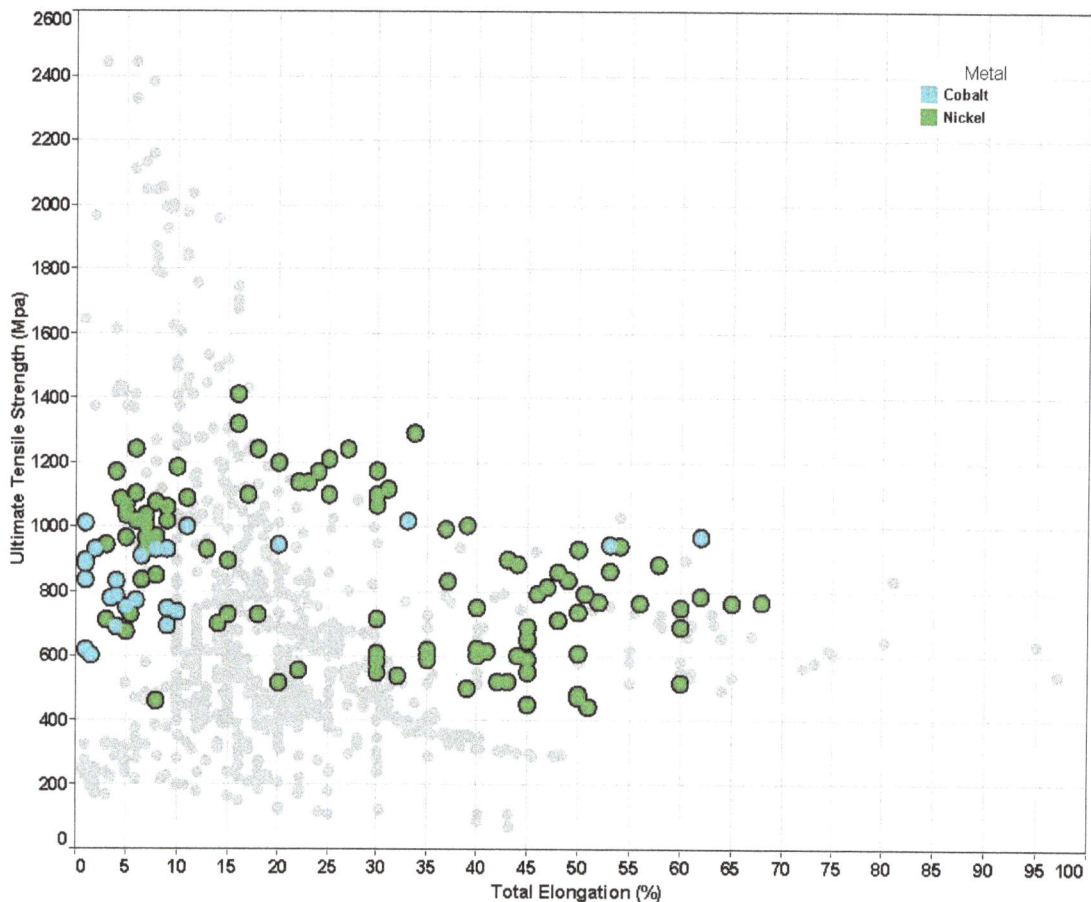

The data shown in Figure 37 were obtained from the Materials Handbook (2008), Davis (1982), and X. Wang et al. (2013).

Tian et al. (2015) showed that, with increasing temperature, the tensile strength of a Ni-based super alloy decreased significantly; however, the yield strength of the material remained constant. Tian et al. also reported that, with increasing temperature, the amount of strain hardening and elongation to failure decreased. In the cases tested at 400°C, the material displayed obvious signs of Portevin-Le Chatelier effect or serrated flow during plastic deformation brought on by the interaction of solute atoms and mobile dislocations.

4.4 Dynamic tensile properties

X. Wang et al. (2013) showed that, with increasing strain rate, the yield strength of Inconel 718 remained the same. However, with increasing strain rate, the amount of ductility obtained was significantly increased. X. Wang et al. also showed that, with increasing temperature, the strength of the material drastically decreased but had no apparent effect on the ductility of the material.

Lee et al. (2011) showed that, when tested in compression at 1000, 3000, and 5000 s⁻¹ and multiple temperatures, the mechanical behavior of Inconel 718 is largely dependent on both temperature and strain rate. The stress in, particular, is very temperature and strain-rate dependent. Inconel 718 exhibits both strain-rate hardening (an increase in stress with increasing strain rate) and temperature softening (a decrease in stress with increasing temperature). Figure 38 shows the stress-strain behavior when tested at 1000 s⁻¹ and four different temperatures. None of the samples tested by Lee et al. failed at any of the given test parameters proving that Ni super alloys display great strength and ductility over a wide range of test conditions.

Figure 38. Inconel 718 stress-strain behavior at strain rates of 1000, 3000, and 5000 s⁻¹ and 4 different temperatures (Lee et al. 2011).

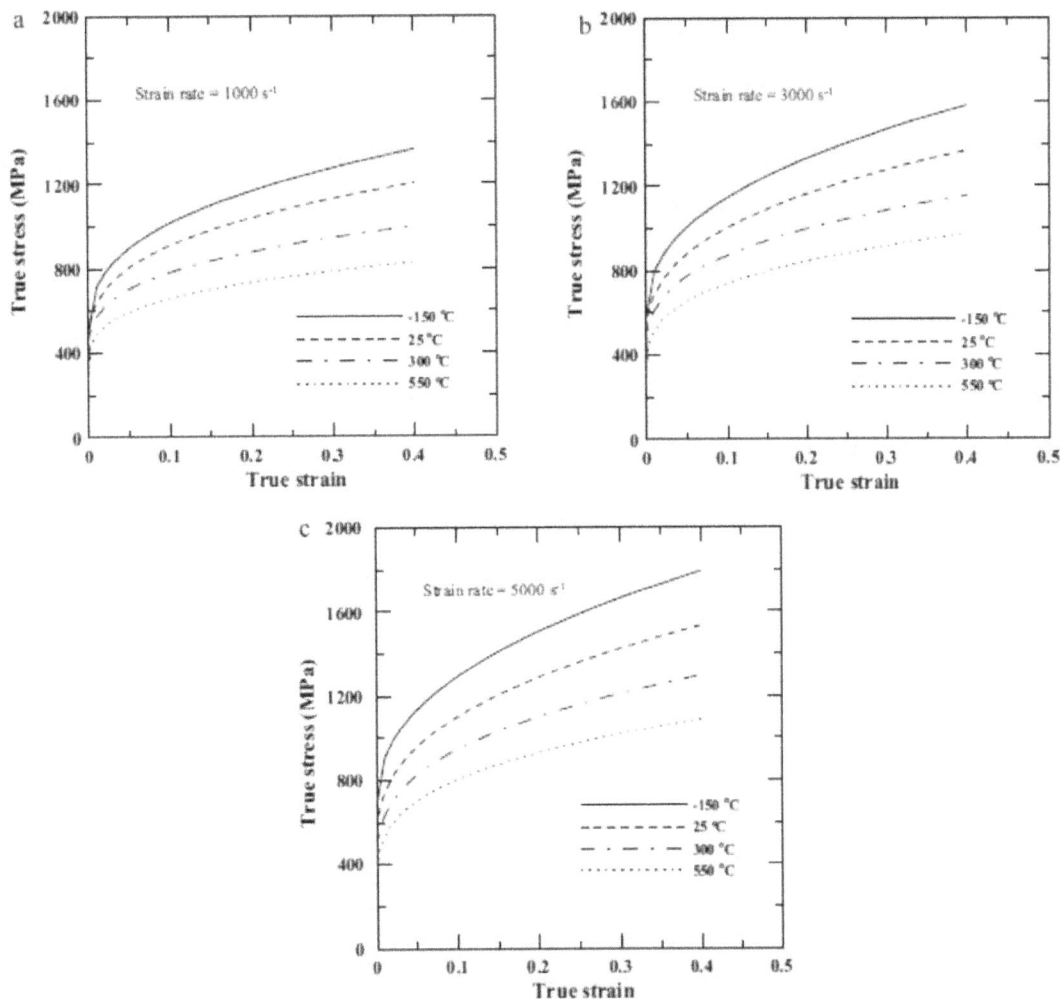

5 Titanium Alloys

5.1 Overview

Titanium metals are comparable in strength and ductility to advanced high-strength steels, low-alloy steels, and high-strength low-alloy steels, but are lighter weight, more corrosion-resistant, and stable at higher temperatures. Due to its high melting point of 1668°C, titanium is termed a refractory metal.

These alloys are commonly used to produce structural components for aircraft, jet engines, and chemical processing equipment. Titanium reportedly is the fifth most commonly used structural metal based on tonnage and the seventh most abundant metallic element on earth.

Titanium consists of two stable allotropes α and β, where α structures are lower in strength and higher in ductility. β structures are very high in strength but occur only in titanium alloys exposed to temperatures greater than 882°C or specifically alloyed and processed to make the β structures stable at lower temperatures. Titanium alloys may contain either or both of these allotropes.

Similar to AHSS, it is possible for mechanical twins to develop or for α structures to transform with heat and deformation into β structures.

5.2 Common designations

In the United States, titanium is usually designated by grade, which is defined by ASTM B 265 and dependent on chemical composition.

The most common titanium alloy is Ti-6Al-4V, a high-strength alloy termed "grade 5" by ASTM B 265 and R56400 in the UNS.

Titanium is commonly alloyed with aluminum, iron, vanadium, zirconium, molybdenum, tin, niobium, chromium, silicon, nickel, and ruthenium (Cardarelli 2008).

5.3   Material properties

Typical material properties for titanium alloys are in Table 8.

Table 8. Material properties of titanium alloys (Cardarelli 2008).

Property	Value
Specific Gravity	4.5-6.2, 4.54 (pure Ti)
Elastic Modulus (GPa)	102 (pure Ti)
Poisson's Ratio (GPa)	.34 (pure Ti)
Coefficient of linear Thermal Expansion $\left(\frac{1}{K} \cdot 10^{-6}\right)$	8.7 (pure Ti)
Electrical Resistivity ρ: $(\mu\Omega \cdot cm)$	56 (pure Ti)

Titanium alloys have excellent corrosion resistance due to a protective passive layer that forms in the presence of oxygen or moisture. These alloys are also lighter weight than steels of comparable strength and ductility (Cardarelli 2008).

5.4   Quasi-static tensile properties

The quasi-static mechanical properties of titanium alloys are comparable to advanced high-strength steels, low-alloy steels, and high-strength low-alloy steels. Figure 39 shows how the mechanical properties of titanium compare to other metals in this report.

Figure 39. Quasi-static ultimate tensile strength and total elongation properties for titanium-based alloys.

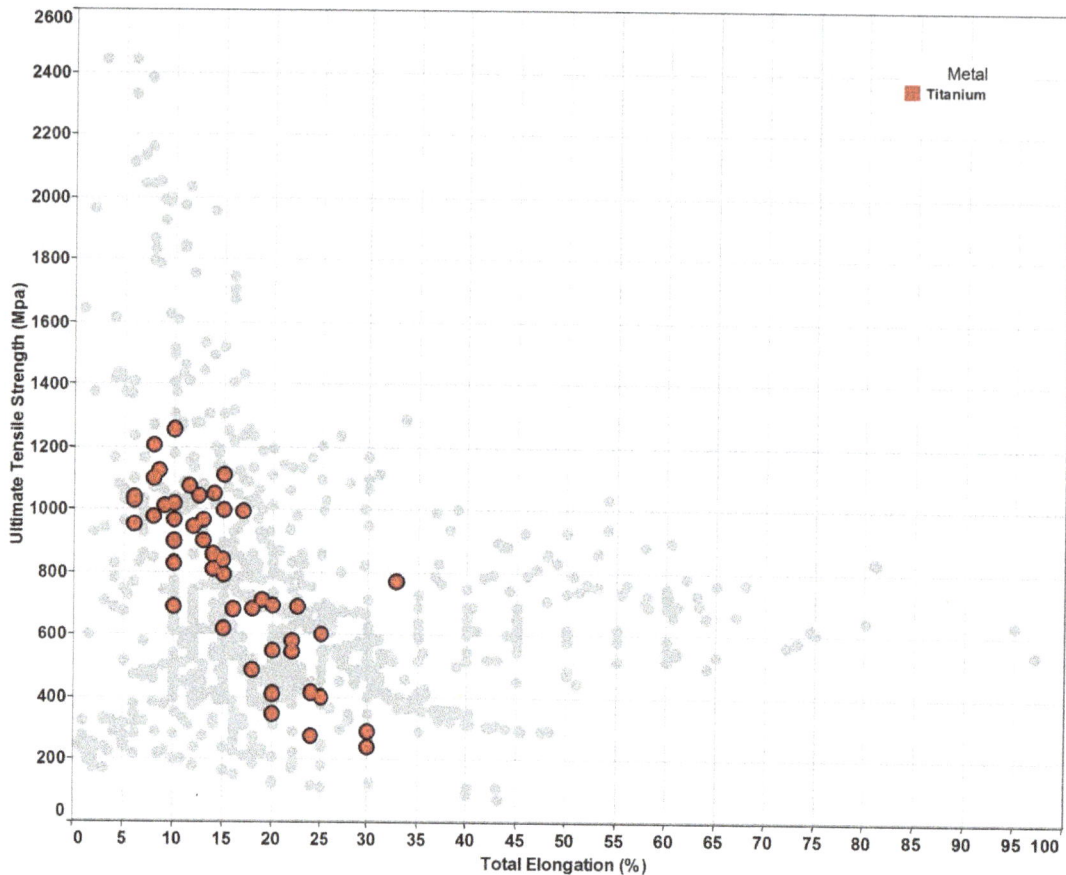

The data shown in Figure 39 were obtained from the Materials Handbook (2008) and the ASM Alloy Center Database™ (2015).

5.5 Dynamic properties

The dynamic mechanical properties of titanium are more focused on compressive strength than other metals. A significant increase in compressive (Zhan et al. 2014a; Zhan et al. 2014b; G. Zhang et al. 2014; Khan et al. 2004; Lee and Lin 1998; Nemat-Nasser et al. 1999; and Chen et al. 2015), tensile (Q. Zhang et al. 2014), and shear (Liao and Duffy 1998) strengths is seen when the strain rate is increased from $10^{-3}\,s^{-1}$ to $10^{3}\,s^{-1}$. A decrease in compressive strength is observed when the material is tested at elevated temperatures (Zhan et al. 2014a; Zhan et al. 2014b; Khan et al. 2004; Lee and Lin 1998; Nemat-Nasser et al. 1999; and Chen et al. 2015). Although the compressive strength is significantly reduced at elevated temperatures, the maximum flow stress in compression is still near 500 MPa at 1100°C for both a Ti-6Cr-5Mo-5V-4Al and a Ti-6Al-4V alloy (Zhan et al. 2014b, and Lee and Lin 1998).

Since many authors did not test the titanium samples to failure, it is difficult to determine if the material experienced greater deformation at high strain-rates. However, Q. Zhang et al. (2014) found that a Ti-6.6Al-3.3Mo-1.8Zr-0.29Si titanium alloy was slightly less ductile under compressive high-rate loading and significantly less ductile under high-rate tensile testing. Q. Zhang et al. (2014) also noted strain-softening behavior during high strain-rate tensile testing and strain-hardening behavior during compression testing at both quasi-static and high strain rates. These results are shown in Figure 40 and Figure 41.

Lee and Lin (1998) found that the Johnson-Cook constitutive model closely fit the experimental results of a Ti-6Al-4V alloy tested at high strain rates under temperatures ranging from 25°C to 1100°C.

Zhan et al. (2014b) found that, although the Johnson-Cook constitutive model was a good fit for the compressive true stress-true strain test results of a Ti-6Cr-5Mo-5V-4Al tested at temperatures ranging from 20°C to 1100°C, the modified Zerilli-Armstrong constitutive model was an even better fit.

Figure 40. Quasi-static and high strain-rate tensile testing of a Ti-6.6Al-3.3Mo-1.8Zr-0.29Si alloy (from Q. Zhang et al. 2014).

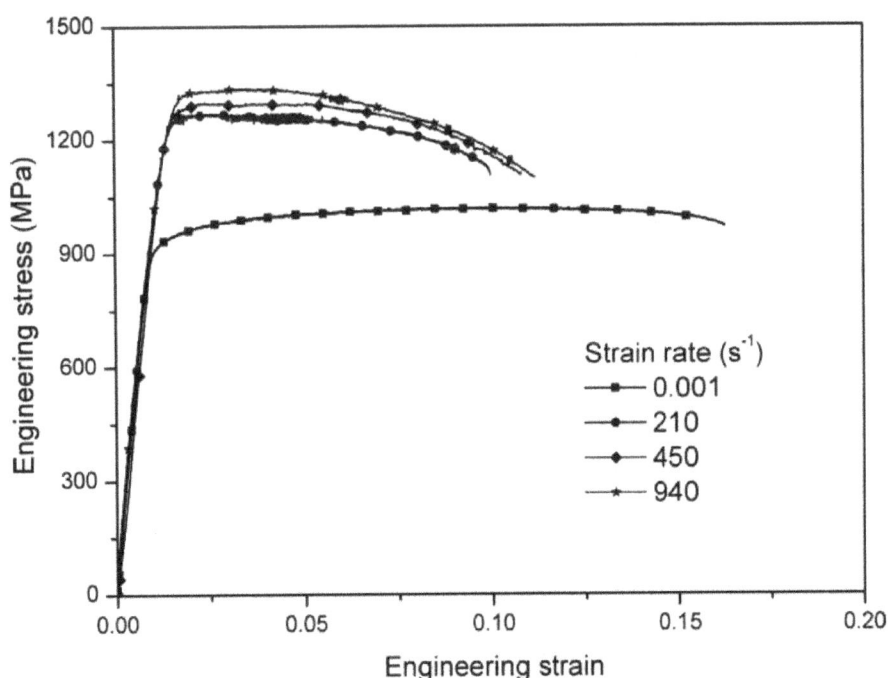

Figure 41. Quasi-static and high strain-rate compression testing of a Ti-6.6Al-3.3Mo-1.8Zr-0.29Si alloy (Q. Zhang et al. 2014).

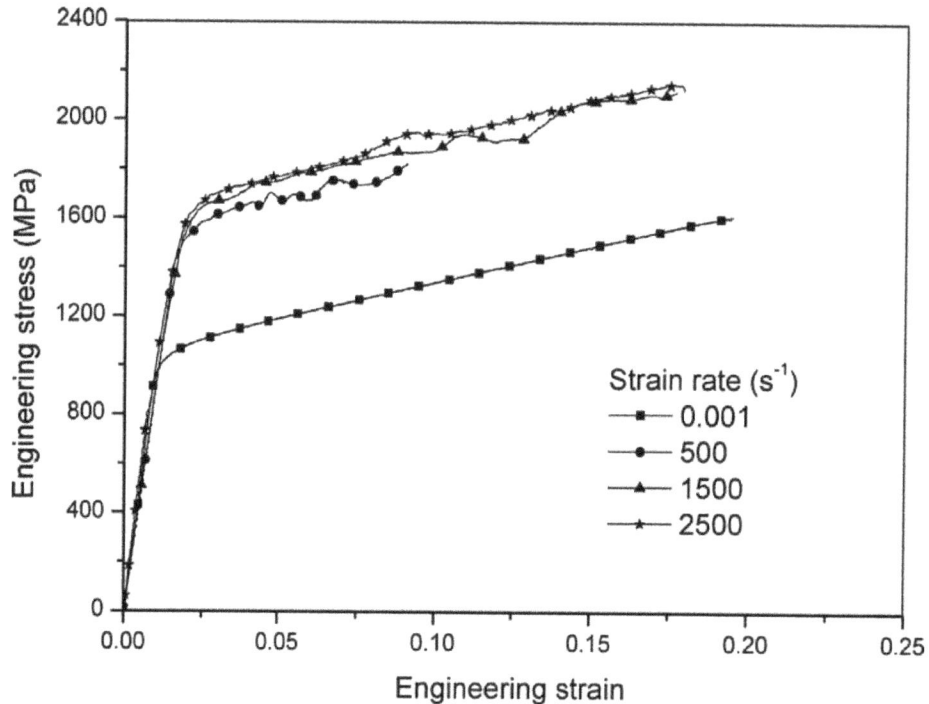

Khan et al. (2004) and Chen et al. (2015) both studied the fit of the Johnson-Cook and Khan-Huang-Liang constitutive models for a Ti-6Al-4V alloy tested at quasi-static and high strain rates and at temperatures between -40°C and 482°C (Khan et al. 2004) / 20°C and 900°C (Chen et al. 2015). Both studies found that the Khan-Huang-Liang model was a better fit than the Johnson-Cook model. However, Chen et al. found that, when the Johnson-Cook and Khan-Huang-Liang models were modified by adding a temperature-dependent work hardening parameter, both methods provided an accurate fit with a standard deviation of 40-45 MPa (less than 10%).

6   Summary

This report reviewed the mechanical properties of iron-, aluminum-, nickel-, cobalt-, and titanium-based metals with the goal of ultimately using these metals as part of a composite armoring system to protect critical bridges from terrorist attacks. An optimized composite armor system will likely include metal alloys with different properties as well as other non-metallic materials.

The optimal properties of each metal alloy can be exploited though proper placement within the armoring system based on the desired protective mechanism and environmental exposure conditions.

Plain carbon and low-alloy steels are widely available and are commonly used in construction. Additionally, many low-alloy and high-strength low-alloy (HSLA) steels are comparable in strength and ductility to many alloys classified as advanced high-strength steel, stainless steel, nickel alloys, cobalt alloys, or titanium.

Advanced high-strength steels (AHSS), like many of the low-alloy and HSLA steels, have good strength and ductility and may be comparable in cost. Experimental data indicate that the strength and ductility of highly martensitic steels are relatively insensitive to strain rate. In addition to having moderate strength, primarily austenitic, twinning induced plasticity (TWIP) steels can withstand large deformations that in some cases may be as high as 90% when loaded in tension. However, while this deformation may help dissipate energy from a collision or blast-type event, this property is useless if the site geometry cannot permit such large deformations.

Stainless steels have comparable mechanical properties to AHSS and undergo similar deformation mechanisms. Stainless steels are reported to be as ductile as TWIP steels. However, the tensile strength of stainless steel is greater than AHSS in scenarios where moderate deformation (20-40%) can occur. Stainless steels also exhibit moderate-to-good corrosion resistance and are more suitable for use in marine environments.

High-performance steels vary in composition and production methods, but they are very high strength (1,000-2,500 MPa in tension). HP alloys have low to moderate ductility (0-20%), and some alloys even have moderate corrosion resistance.

Aluminum alloys are much lower in strength than the other metals studied in this report and have comparable or lower ductility. However, these alloys are light weight and easily machined, and many alloys are an economical alternative to steel. The low strength of aluminum may be beneficial if used as part of a multi-layered composite system, as long as it does not react with the other materials used.

Nickel- and cobalt-based alloys are moderate in strength and have good corrosion resistance. Some nickel alloys can withstand deformations as high as 60% without a significant reduction in strength. Many nickel and cobalt alloys also have moderate to high tensile strength (800-1,200 MPa) and relatively low ductility, which may be beneficial for some armoring systems.

Titanium alloys have mechanical properties similar to AHSS, low-alloy steels, and HSLA steels, but also are light weight, stable at higher temperatures than other metals, and highly corrosion-resistant.

In the end, the selection of the most appropriate metal(s) for a composite armoring system is dependent on the design threat characteristics, compatibility with other armoring system materials, and site criteria.

References

Amato, K. N., S. M. Gaytan, L. E. Murr, E. Martinez, P. W. Shindo, J. Hernandez, S. Collins, F. Medina. 2012. Microstructures and mechanical behavior of Inconel 718 fabricated by selective laser melting. *Acta Materialia* 60(5):2229-2239. doi:10.1016/j.actamat.2011.12.032

ASM Alloy Center Database™. 2015. Materials properties. http://mio.asminternational.org/ac/.

ASM Handbook. 1991. *Volume 4: heat treating.* ASM, 377. Novelty, OH: ASM International.

ASTM International. 1998. *Standard specification for steel, sheet and strip, high-strength, low-alloy, columbium or vanadium, or both, hot-rolled and cold-rolled (withdrawn 2000).* Designation A607-98. Philadelphia, PA: ASTM International.

_____. 2009. *Standard specification for steel, sheet and strip, high-strength, low-alloy, hot-rolled and cold-rolled, with improved atmospheric corrosion resistance.* Designation A606/A606M-09a. Philadelphia, PA: ASTM International.

_____. 2012. *Standard specification for pressure vessel plates, carbon steel, low- and intermediate-tensile strength.* Designation A285/A285M-12. Philadelphia, PA: ASTM International.

_____. 2013. *Standard specification for high-strength low-alloy nickel, copper, phosphorus steel H-piles and sheet piling with atmospheric corrosion resistance for use in marine environments.* Designation A690/A690M-13a. Philadelphia, PA: ASTM International.

_____. 2013. *Standard specification for high-strength low-alloy structural steel.* Designation A242/A 242M-13. Philadelphia, PA: ASTM International.

_____. 2013. *Standard specification for hot-rolled and cold-finished age-hardening stainless steel bars and shapes.* Designation A564/A564M-13. Philadelphia, PA: ASTM International.

_____. 2013. *Standard specification for hot-rolled structural steel, high-strength low-alloy plate with improved formability.* Designation A656/A656M-13. Philadelphia, PA: ASTM International.

_____. 2013. *Standard specification for low and intermediate tensile strength carbon steel plates.* Designation A283/A283M-13. Philadelphia, PA: ASTM International.

_____. 2013. *Standard specification for normalized high-strength low-alloy structural steel plates.* Designation A633/A633M-13. Philadelphia, PA: ASTM International.

_____. 2013. *Standard specification for precipitation–strengthened low-carbon nickel-copper-chromium-molybdenum-columbium alloy structural steel plates.* Designation A710/A710M-02(2013). Philadelphia, PA: ASTM International.

_____. 2013. *Standard specification for steel bars, carbon, merchant quality, m-grades.* Designation A575-96(2013)e1. Philadelphia, PA: ASTM International.

_____. 2013. *Standard specification for steel, sheet, cold-rolled, complex phase (CP), dual phase (DP) and transformation induced plasticity (TRIP).* Designation A1088-13. Philadelphia, PA: ASTM International.

_____. 2013. *Standard specification for structural carbon steel plates of improved toughness.* Designation A573/A573M-13. Philadelphia, PA: ASTM International.

_____. 2013. *Standard specification for structural steel for bridges.* Designation A709/A709M-13a. Philadelphia, PA: ASTM International.

_____. 2013. *Standard specification for titanium and titanium alloy strip, sheet, and plate.* Designation B265-13ae1. Philadelphia, PA: ASTM International.

_____. 2014. *Standard specification for carbon structural steel.* Designation A36/A36M-14. Philadelphia, PA: ASTM International.

_____. 2014. *Standard specification for cold-formed welded and seamless high-strength, low-alloy structural tubing with improved atmospheric corrosion resistance.* Designation A847/A847M-14. Philadelphia, PA: ASTM International.

_____. 2014. *Standard specification for high-strength carbon-manganese steel of structural quality.* Designation A529/A529M-14. Philadelphia, PA: ASTM International.

_____. 2014. *Standard specification for high-strength low-alloy structural steel plate with atmospheric corrosion resistance.* Designation A871/A871M-14. Philadelphia, PA: ASTM International.

_____. 2014. *Standard specification for high-yield-strength, quenched and tempered alloy steel plate, suitable for welding.* Designation A514/A514M-14. Philadelphia, PA: ASTM International.

_____. 2014. *Standard specification for structural steel for ships.* Designation A131/A131M-14. Philadelphia, PA: ASTM International.

_____. 2014. Standard specification for structural steel with improved yield strength at high temperature for use in buildings. Designation A1077 / A1077M-14. Philadelphia, PA: ASTM International.

_____. 2015. *Standard specification for high-strength low-alloy columbium-vanadium structural steel.* Designation A572/A572M-15. Philadelphia, PA: ASTM International.

_____. 2015. *Standard specification for high-strength low-alloy steel shapes of structural quality, produced by quenching and self-tempering process (QST).* Designation A913/A913M-15. Philadelphia, PA: ASTM International.

_____. 2015. *Standard specification for high-strength low-alloy structural steel, up to 50 ksi [345 MPa] minimum yield point, with atmospheric corrosion resistance.* Designation A588/A588M-15. Philadelphia, PA: ASTM International.

_____. 2015. *Standard specification for hot-formed welded and seamless high-strength low-alloy structural tubing.* Designation A618/A618M-04(2015). Philadelphia, PA: ASTM International.

_____. 2015. *Standard specification for structural steel shapes.* Designation A992/A992M-11(2015). Philadelphia, PA: ASTM International.

Bardelcik, A., C. P Salisbury, S. Winkler, M. A. Wells, and M. J. Worswick. 2010. Effect of cooling rate on the high strain rate properties of boron steel. *International Journal of Impact Engineering* 37(6):694-702. doi:http://dx.doi.org/10.1016/j.ijimpeng.2009.05.009

Bardelcik, A., M. J. Worswick, S. Winkler, and M. A. Wells. 2012. A strain rate sensitive constitutive model for quenched boron steel with tailored properties. *International Journal of Impact Engineering* 50(0):49-62. doi:http://dx.doi.org/10.1016/j.ijimpeng.2012.06.007

Billur, M., and T. Altan. 2012. Challenges in forming advanced high strength steels. In *Proceedings of New Developments in Sheet Metal Forming, 2-4 May*, 285-304. Stuttgart, Germany.

Boh, J. W., L. A. Louca, and Y. S. Choo. 2004. Strain rate effects on the response of stainless steel corrugated firewalls subjected to hydrocarbon explosions. *Journal of Constructional Steel Research* 60:1-29. doi:10.1016/j.jcsr.2003.08.005

Bouaziz, O., and D. Barbier. 2013. Benefits of recovery and partial recrystallization of nano-twinned austenitic steels. *Advanced Engineering Materials* 15:976-979. doi:10.1002/adem.201300045

Byun, T., N. Hashimoto, and K. Farrell. 2004. Temperature dependence of strain hardening and plastic instability behaviors in austenitic stainless steels. *Acta Materialia* 52(13):3889-3899.

Byun, T. S., and K. Farrell. 2003. Tensile properties of Inconel 718 after low temperature neutron irradiation. *Journal of Nuclear Materials* 318:292-299. doi:10.1016/s0022-3115(03)00006-0

Cardarelli, F. 2008. *Ferrous metals and their alloys.* ASM International, 59-157. London: Springer.

Chandler, H. 1994. *Heat treater's guide: practices and procedures for irons and steels.* Novelty, OH: ASM International.

Chen, G., C. Ren, X. Qin, and J. Li. 2015. Temperature dependent work hardening in Ti–6Al–4V alloy over large temperature and strain rate ranges: Experiments and constitutive modeling. *Materials and Design* 83:598-610. doi:http://dx.doi.org/10.1016/j.matdes.2015.06.048

Chen, Y., A. H. Clausen, O. S. Hopperstad, and M. Langseth. 2009. Stress–strain behaviour of aluminium alloys at a wide range of strain rates. *International Journal of Solids and Structures* 46(21):3825-3835. doi:http://dx.doi.org/10.1016/j.ijsolstr.2009.07.013

Chlebus, E., K. Gruber, B. Kuznicka, J. Kurzac, and T. Kurzynowski. 2015. Effect of heat treatment on the microstructure and mechanical properties of Inconel 718 processed by selective laser melting. *Materials Science and Engineering a-Structural Materials Properties Microstructure and Processing* 639:647-655. doi:10.1016/j.msea.2015.05.035

Curtze, S., and V. T. Kuokkala. 2010. Dependence of tensile deformation behavior of TWIP steels on stacking fault energy, temperature and strain rate. *Acta Materialia* 58(15):5129-5141. doi:http://dx.doi.org/10.1016/j.actamat.2010.05.049

Das, A., S. Sivaprasad, P. C. Chakraborti, and S. Tarafder. 2011. Morphologies and characteristics of deformation induced martensite during low cycle fatigue behaviour of austenitic stainless steel. *Materials Science and Engineering A* 528(27):7909-7914. doi:http://dx.doi.org/10.1016/j.msea.2011.07.011

Das, A., S. Sivaprasad, M. Ghosh, P. C. Chakraborti, and S. Tarafder. 2008. Morphologies and characteristics of deformation induced martensite during tensile deformation of 304 LN stainless steel. *Materials Science and Engineering A* 486(1–2):283-286. doi:http://dx.doi.org/10.1016/j.msea.2007.09.005

Davis, J. R. 1998. *Metals handbook desk edition, 2nd edition.* Novelty, OH: ASM International.

Davis, J. R. 2001. *Alloying: understanding the basics.* Novelty, OH: ASM International.

De, A. K., J. G. Speer, D. K. Matlock, D. C. Murdock, M. C. Mataya, and R. J. Comstock Jr. 2006. Deformation-induced phase transformation and strain hardening in type 304 austenitic stainless steel. *Metallurgical and Materials Transactions A* 37(6):1875-1886.

Ding, H., Z.-Y. Tang, W. Li, M. Wang, and D. Song. 2006. Microstructures and Mechanical Properties of Fe-Mn-(Al, Si) TRIP/TWIP Steels. *Journal of Iron and Steel Research, International* 13(6):66-70. doi:http://dx.doi.org/10.1016/S1006-706X(06)60113-1

Dini, G., A. Najafizadeh, R. Ueji, and S. M. Monir-Vaghefi. 2010. Improved tensile properties of partially recrystallized submicron grained TWIP steel. *Materials Letters* 64:15-18. doi:10.1016/j.matlet.2009.09.057

Frommeyer, G., U. Brüx, and P. Neumann. 2003. Supra-ductile and high-strength manganese-TRIP/TWIP steels for high energy absorption purposes. *ISIJ International* 43(3):438-446. doi:10.2355/isijinternational.43.438

Ghosh, A., A. Adesola, J. A. Szpunar, A. G. Odeshi, and N. P. Gurao. 2015. Effect of tempering conditions on dynamic deformation behaviour of an aluminium-lithium alloy. *Materials & Design* 81:1-10. doi:10.1016/j.matdes.2015.04.050

Grässel, O., L. Krüger, G. Frommeyer, and L. W. Meyer. 2000. High strength Fe–Mn–(Al, Si) TRIP/TWIP steels development — properties - application. *International Journal of Plasticity* 16(10–11):1391-1409. doi:http://dx.doi.org/10.1016/S0749-6419(00)00015-2

Guan, K., Z. Wang, M. Gao, X. Li, and X. Zeng. 2013. Effects of processing parameters on tensile properties of selective laser melted 304 stainless steel. *Materials and Design* 50(0):581-586. doi:http://dx.doi.org/10.1016/j.matdes.2013.03.056

Guo, W.-G., and S. Nemat-Nasser. 2006. Flow stress of Nitronic-50 stainless steel over a wide range of strain rates and temperatures. *Mechanics of Materials* 38(11):1090-1103. doi:http://dx.doi.org/10.1016/j.mechmat.2006.01.004

Guo, W.- G., X. Q. Zhang, J. Su, Y. Su, Z. Y. Zeng, and X. J. Shao. 2011. The characteristics of plastic flow and a physically-based model for 3003 Al–Mn alloy upon a wide range of strain rates and temperatures. *European Journal of Mechanics - A/Solids* 30(1):54-62. doi:http://dx.doi.org/10.1016/j.euromechsol.2010.09.001

Hadianfard, M. J., R. Smerd, S. Winkler, and M. Worswick. 2008. Effects of strain rate on mechanical properties and failure mechanism of structural Al–Mg alloys. *Materials Science and Engineering A* 492(1–2):283-292. doi:http://dx.doi.org/10.1016/j.msea.2008.03.037

Hamada, A. S., L. P. Karjalainen, R. D. K. Misra, and J. Talonen. 2013. Contribution of deformation mechanisms to strength and ductility in two Cr-Mn grade austenitic stainless steels. *Materials Science and Engineering a-Structural Materials Properties Microstructure and Processing* 559:336-344. doi:10.1016/j.msea.2012.08.108

Harding, J., E. Wood, and J. Campbell. 1960. Tensile testing of materials at impact rates of strain. *Journal of Mechanical Engineering Science* 2:88-96.

Harvey, P. D. 1982. *Engineering properties of steel*. Novelty, OH: American Society for Materials.

Huh, H., S.-B. Kim, J.-H. Song, and J.-H. Lim. 2008. Dynamic tensile characteristics of TRIP-type and DP-type steel sheets for an auto-body. *International Journal of Mechanical Sciences* 50(5):918-931. doi:http://dx.doi.org/10.1016/j.ijmecsci.2007.09.004

Institute, S. M. D. 2014. AHSS data utilization. http://www.autosteel.org/research/ahss-data-utilization.aspx.

Kanagarajah, P., F. Brenne, T. Niendorf, and H. J. Maier. 2013. Inconel 939 processed by selective laser melting: Effect of microstructure and temperature on the mechanical properties under static and cyclic loading. *Materials Science and Engineering a-Structural Materials Properties Microstructure and Processing* 588:188-195. doi:10.1016/j.msea.2013.09.025

Keeler, S. 2014. Advanced high-strength steels application guidelines. In *World Auto Steel*, ed. M. Kimchi. Brussels: World Auto Steel.

Khan, A. S., Y. Sung Suh, and R. Kazmi. 2004. Quasi-static and dynamic loading responses and constitutive modeling of titanium alloys. *International Journal of Plasticity* 20(12):2233-2248. doi:http://dx.doi.org/10.1016/j.ijplas.2003.06.005

Khosravifard, A., M. M. Moshksar, and R. Ebrahimi. 2013. High strain rate torsional testing of a high manganese steel: Design and simulation. *Materials and Design* 52:495-503. doi:10.1016/j.matdes.2013.05.083

Kim, J.-H., D. Kim, H. N. Han, F. Barlat, and M.-G. Lee. 2013. Strain rate dependent tensile behavior of advanced high strength steels: Experiment and constitutive modeling. *Materials Science and Engineering A* 559(0):222-231. doi:http://dx.doi.org/10.1016/j.msea.2012.08.087

Kolsky, H. 1949. An investigation of the mechanical properties of materials at very high rates of loading. *Proceedings of the Physical Society* 62B(11):676.

Kozmel, T., M. Vural, and S. Tin. 2015. EBSD analysis of high strain rate application Al–Cu based alloys. *Materials Science and Engineering A* 630:99-106. doi:http://dx.doi.org/10.1016/j.msea.2015.02.024

Lee, W.-S., and C.-F Lin. 1998. Plastic deformation and fracture behaviour of Ti–6Al–4V alloy loaded with high strain rate under various temperatures. *Materials Science and Engineering A* 241(1–2):48-59. doi:http://dx.doi.org/10.1016/S0921-5093(97)00471-1

Lee, W.-S., C.-F. Lin, T.-H. Chen, and H.-W. Chen. 2011. Dynamic mechanical behaviour and dislocation substructure evolution of Inconel 718 over wide temperature range. *Materials Science and Engineering a-Structural Materials Properties Microstructure and Processing* 528(19-20):6279-6286. doi:10.1016/j.msea.2011.04.079

Lee, W.-S., C.-F. Lin, T.-H. Chen, and M. C. Yang. 2010. High temperature microstructural evolution of 304L stainless steel as function of pre-strain and strain rate. *Materials Science and Engineering a-Structural Materials Properties Microstructure and Processing* 527:3127-3137. doi:10.1016/j.msea.2010.02.007

Liao, S.-c., and J. Duffy. 1998. Adiabatic shear bands in a TI-6Al-4V titanium alloy. *Journal of the Mechanics and Physics of Solids* 46(11):2201-2231. doi:http://dx.doi.org/10.1016/S0022-5096(98)00044-1

Melby, E. A., and H. O. S. Eide. 2013. Blast loaded aluminium plates: experiments and numerical simulations. Trondheim, Norway: Norwegian University of Science and Technology.

Milad, M., N. Zreiba, F. Elhalouani, and C. Baradai. 2008. The effect of cold work on structure and properties of AISI 304 stainless steel. *Journal of Materials Processing Technology* 203(1):80-85.

Mukai, T., K. Ishikawa, and K. Higashi. 1995. Influence of strain rate on the mechanical properties in fine-grained aluminum alloys. *Materials Science and Engineering A* 204(1–2):12-18. doi:http://dx.doi.org/10.1016/0921-5093(95)09929-8

Nemat-Nasser, S., W. G. Guo, and J. Y. Cheng. 1999. Mechanical properties and deformation mechanisms of a commercially pure titanium. *Acta Materialia* 47(13):3705-3720. doi:http://dx.doi.org/10.1016/S1359-6454(99)00203-7

Newson, T., and A. P. O. Abp. 2001. Stainless steels-applications, grades, and human exposure. Brussels, Belgium: Eurofer Stainless Steels Production Group.

Ramesh, K. 2008. High rates and impact experiments. In *Springer handbook of experimental solid mechanics*, ed. W. N. Sharpe Jr., 929-960. New York: Springer US.

Sadagopan, S., and D. Urban. 2003. *Formability characterization of a new generation of high strength steels*. AISI/DOE technology roadmap program. Washington, DC: American Iron and Steel Institute.

Sahu, P., S. Curtze, A Das, B. Mahato, V.-T. Kuokkala, and S. G. Chowdhury. 2010. Stability of austenite and quasi-adiabatic heating during high-strain-rate deformation of twinning-induced plasticity steels. *Scripta Materialia* 62(1):5-8. doi:http://dx.doi.org/10.1016/j.scriptamat.2009.09.010

Smerd, R., S. Winkler, C. Salisbury, M. Worswick, D. Lloyd, and M. Finn. 2005. High strain rate tensile testing of automotive aluminum alloy sheet. *International Journal of Impact Engineering* 32(1–4):541-560. doi:http://dx.doi.org/10.1016/j.ijimpeng.2005.04.013

Society of Automated Engineers. 2008. *Steel, high strength, hot rolled sheet and strip, cold rolled sheet, and coated sheet*. SAE J1392. Warrendale, PA: SAE International.

Tavares, S., P. Pedroza, J. Teodosio, and T. Gurova. 1999. Mechanical properties of a quenched and tempered dual phase steel. *Scripta Materialia* 40(8):887-892.

Thakur, D. G., B. Ramamoorthy, and L. Vijayaraghavan. 2009. Machinability investigation of Inconel 718 in high-speed turning. *The International Journal of Advanced Manufacturing Technology* 45(5):421-429. doi:https://doi.org/10.1007/s00170-009-1987-x

Tian, C., G. Han, C. Cui, and X. Sun. 2015. Effects of Co content on tensile properties and deformation behaviors of Ni-based disk superalloys at different temperatures. *Materials and Design* 88:123-131. doi:http://dx.doi.org/10.1016/j.matdes.2015.08.114

Tucker, M. T., M. F. Horstemeyer, W. R. Whittington, K. N. Solanki, and P. M. Gullett. 2010. The effect of varying strain rates and stress states on the plasticity, damage, and fracture of aluminum alloys. *Mechanics of Materials* 42(10):895-907. doi:http://dx.doi.org/10.1016/j.mechmat.2010.07.003

Wang, W., M. Li, C. He, X. Wei, D. Wang, and H. Du. 2013. Experimental study on high strain rate behavior of high strength 600–1000 MPa dual phase steels and 1200 MPa fully martensitic steels. *Materials and Design* 47(0):510-521. doi:http://dx.doi.org/10.1016/j.matdes.2012.12.068

Wang, X., C. Huang, B. Zou, H. Liu, H. Zhu, and J. Wang. 2013. Dynamic behavior and a modified Johnson-Cook constitutive model of Inconel 718 at high strain rate and elevated temperature. *Materials Science and Engineering a-Structural Materials Properties Microstructure and Processing* 580:385-390. doi:10.1016/j.msea.2013.05.062

Wong, C. 2005. IISI-AutoCo Round-Robin Dynamic Tensile Testing Project. International Iron and Steel Institute. www.worldautosteel.org.

Zhan, H., D. Kent, G. Wang, and M. S. Dargusch. 2014a. The dynamic response of a beta titanium alloy to high strain rates and elevated temperatures. *Materials Science and Engineering a-Structural Materials Properties Microstructure and Processing* 607:417-426. doi:10.1016/j.msea.2014.04.028

Zhan, H., G. Wang, D. Kent, and M. Dargusch. 2014b. Constitutive modelling of the flow behaviour of a beta titanium alloy at high strain rates and elevated temperatures using the Johnson-Cook and modified Zerilli-Armstrong models. *Materials Science and Engineering a-Structural Materials Properties Microstructure and Processing* 612:71-79. doi:10.1016/j.msea.2014.06.030

Zhang, Q., J. Zhang, and Y. Wang. 2014. Effect of strain rate on the tension–compression asymmetric responses of Ti–6.6Al–3.3Mo–1.8Zr–0.29Si. *Materials and Design* 61:281-285. doi:http://dx.doi.org/10.1016/j.matdes.2014.05.004

Zhang, X., H. Deng, S. Xiao, Z. Zhang, J. Tang, L. Deng, and W. Hu. 2014. Diffusion of Co, Ru and Re in Ni-based superalloys: A first-principles study. *Journal of Alloys and Compounds* 588:163-169. doi:10.1016/j.jallcom.2013.11.024

REPORT DOCUMENTATION PAGE		Form Approved OMB No. 0704-0188

Public reporting burden for this collection of information is estimated to average 1 hour per response, including the time for reviewing instructions, searching existing data sources, gathering and maintaining the data needed, and completing and reviewing this collection of information. Send comments regarding this burden estimate or any other aspect of this collection of information, including suggestions for reducing this burden to Department of Defense, Washington Headquarters Services, Directorate for Information Operations and Reports (0704-0188), 1215 Jefferson Davis Highway, Suite 1204, Arlington, VA 22202-4302. Respondents should be aware that notwithstanding any other provision of law, no person shall be subject to any penalty for failing to comply with a collection of information if it does not display a currently valid OMB control number. **PLEASE DO NOT RETURN YOUR FORM TO THE ABOVE ADDRESS.**

1. REPORT DATE (DD-MM-YYYY) January 2023	2. REPORT TYPE Final	3. DATES COVERED (From - To)

4. TITLE AND SUBTITLE	5a. CONTRACT NUMBER
State-of-Practice on the Mechanical Properties of Metals for Armor-Plating	5b. GRANT NUMBER
	5c. PROGRAM ELEMENT NUMBER

6. AUTHOR(S)	5d. PROJECT NUMBER
Wendy R. Long, Zackery B. McClelland, Dylan A. Scott, and C. Kennan Crane	5e. TASK NUMBER
	5f. WORK UNIT NUMBER

7. PERFORMING ORGANIZATION NAME(S) AND ADDRESS(ES)	8. PERFORMING ORGANIZATION REPORT NUMBER
Geotechnical and Structures Laboratory US Army Engineer Research and Development Center 3909 Halls Ferry Road Vicksburg, MS 39180-6199	ERDC/GSL TR-23-2

9. SPONSORING / MONITORING AGENCY NAME(S) AND ADDRESS(ES)	10. SPONSOR/MONITOR'S ACRONYM(S)
Federal Highway Administration Turner-Fairbank Highway Research Center McLean, VA 22101	11. SPONSOR/MONITOR'S REPORT NUMBER(S)

12. DISTRIBUTION / AVAILABILITY STATEMENT
Approved for public release; distribution is unlimited.

13. SUPPLEMENTARY NOTES

Material Specifications for Attack Countermeasures on Bridges, IAA DTFH61-10-X-30028 and IAA DTFH61-13-X-30049

14. ABSTRACT

This report presents a review of quasi-static and dynamic properties of various iron, titanium, nickel, cobalt, and aluminum metals. The physical and mechanical properties of these materials are crucial for developing composite armoring systems vital for protecting critical bridges from terrorist attacks. When the wide range of properties these materials encompass is considered, it is possible to exploit the optimal properties of metal alloys though proper placement within the armoring system, governed by desired protective mechanism and environmental exposure conditions.

15. SUBJECT TERMS		
Mechanical properties	High-strain-rate loading	Cobalt
Quasi-static loading	Armor plating	Aluminum
Terrorism	Iron	Iron and steel bridges–Blast effect
Armor-plate	Steel	Iron and steel bridges–Explosions
	Titanium	Iron and steel bridges–Protection
	Nickel	

16. SECURITY CLASSIFICATION OF:			17. LIMITATION OF ABSTRACT	18. NUMBER OF PAGES	19a. NAME OF RESPONSIBLE PERSON
a. REPORT Unclassified	b. ABSTRACT Unclassified	c. THIS PAGE Unclassified	SAR	87	19b. TELEPHONE NUMBER (include area code)

BODY-87

Standard Form 298 (Rev. 8-98)
Prescribed by ANSI Std. 239.18

www.ingramcontent.com/pod-product-compliance
Lightning Source LLC
Chambersburg PA
CBHW081509200326
41518CB00015B/2435